T0299361

Intrinsic Capability

Implementing Intrinsic Sustainable Development for an Ecological Civilisation

Intrinsic Capability

Implementing Intrinsic Sustainable Development for an Ecological Civilisation

editors

Frank Birkin
Sheffield University, UK

Thomas Polesie
Gothenburg University, Sweden

W⊖ World Scientific

NEW JERSEY · LONDON · SINGAPORE · BEIJING · SHANGHAI · HONG KONG · TAIPEI · CHENNAI · TOKYO

Published by

World Scientific Publishing Co. Pte. Ltd.

5 Toh Tuck Link, Singapore 596224

USA office: 27 Warren Street, Suite 401-402, Hackensack, NJ 07601

UK office: 57 Shelton Street, Covent Garden, London WC2H 9HE

Library of Congress Cataloging-in-Publication Data
Names: Birkin, Frank, editor. | Polesie, Thomas, editor.
Title: Intrinsic capability : implementing intrinsic sustainable development
    for an ecological civilisation / edited by Frank Birkin (Sheffield
    University, UK) and Thomas Polesie (Gothenburg University, Sweden).
Description: New Jersey : World Scientific, [2018]
Identifiers: LCCN 2018055913 | ISBN 9789813225572 (hc : alk. paper)
Subjects: LCSH: Sustainable development.
Classification: LCC HC79.E5 I645 2018 | DDC 338.9/27--dc23
LC record available at https://lccn.loc.gov/2018055913

British Library Cataloguing-in-Publication Data
A catalogue record for this book is available from the British Library.

For any available supplementary material, please visit
https://www.worldscientific.com/worldscibooks/10.1142/10586#t=suppl

Desk Editors: Vimal Thangavel/Sylvia Koh

Typeset by Stallion Press
Email: enquiries@stallionpress.com

Printed in Singapore

Preface

I was preoccupied with three questions while editing this book. The first question is not challenging but needs to be asked — "Do civilisations change?" It is not challenging because our age is so well informed: we now know many ancient civilisations that have come and gone; we can travel and see for ourselves how different people did and do manage to organise themselves into large, civilised collectives; and, when you have lived to the old age shared by both editors of this book, you can appreciative just how your own civilisation has changed over a single lifetime. Civilisations do change.

The second question is "Does our civilisation need to change?" This can be a far more contentious question because we can argue about (i) the nature of "our civilisation" and (ii) the value judgements that help assess any need — or lack of need — for change. I think my personal answer to this question is best expressed by the maxim that the only constant is change itself. We live in a dynamic, evolving universe in which change occurs constantly from quantum to cosmological scales and within which humanity plans, acts, achieves and regrets with varying levels of ignorance.

The third question provides the real guide to this book — "How do civilisations change?" Many texts will identify ideas or key historical figures whose intellect, courage and vision have shaped past civilisations. But is this really the case? Do outstanding ideas or prominent individuals have such an influence? Or should we side with Tolstoy, for example, who argues in his epic novel *War and Peace* (1869) that neither Alexander I nor Napoleon Bonaparte were in control of how their respective civilisations, the Tsarist autocracy of Russia and the first empire of France, developed but were merely swept along by the tides of their times. In either case, the key concepts or prominent people versus tidal swells, there is a need to reflect on our own times. Civilisation is changing now but how, in what way and where is it heading? There are answers in this book but they are neither definitive nor prescriptive; at their best, they will put civilisation change on your agenda and help you make up your own mind.

Frank Birkin
Hammerfest, Norway
13th January 2019

About the Editors

 **Frank Birkin** is a Professor for Accounting and Sustainable Development at the Management School of the University of Sheffield, UK. During his career, he has worked at the inception of key sustainability techniques and approaches that include social and environmental accounting, ecological accounting, cleaner technology and environmental management. He is a co-author of the bestselling book *Intrinsic Sustainable Development: Epistemes, Science, Business and Sustainability* (2012, World Scientific Press) and a founding member of the *Intrinsic Earth* initiative (www.intrinsicearth.org). Frank's current interests focus upon the mobilisation of societies for greater sustainability by using and integrating the fresh insights provided by empirical science with aspects of cultures and traditions.

Thomas Polesie is an Emeritus Professor of Accounting and Finance, University of Gothenburg, School of Business, Economics and Law, Sweden. He has developed a unique approach to accounting that he has applied in the analyses of many companies and sectors, including energy, shipping, the built environment and financial institutions. Thomas also specialises in studies of judgement and evaluation. He is the author of many books and has co-authored seminal publications with Frank Birkin on the topics of sustainable business and episteme change. His current interests also include the implications of current East to West relations and accounting for emerging eco-civilisations.

Contents

Chapter 1

INTRODUCTION

Frank Birkin[*,‡] and Thomas Polesie[†,§]

*University of Sheffield, Sheffield S10 2TN, UK
†University of Gothenburg, 405 30 Gothenburg, Sweden
‡f.birkin@sheffield.ac.uk
§thomas.polesie@handels.gu.se

This book is about a global experiment conducted with two aims. The first aim relates to a new approach to solving global problems such as poverty, pollution, habitat loss and climate change. The second aim is to bring about better lives for all.

The method of the experiment may be expressed figuratively as a move to another planet whilst keeping our feet placed firmly on Planet Earth. This is achieved by a change in our understanding of what makes knowledge possible — we effectively change our knowledge to live in a different world. This change is conceptually simple but what is harder to grasp is the impact of its consequences. The new possibility for knowledge is based on the most detailed and precise science about how we, our planet and the universe function and its consequences penetrate all quarters of life. Mankind has made an enormous investment to acquire this knowledge and we now need to pay it due attention.

Many people are now paying due attention, but they are diverse and unconnected. So in this experiment, we asked them what they understand by the new possibility of knowledge and what do they see as its consequences. Responses have come from all around the world and from many different fields: ecological business, ecological accounting and marketing; traditions and sustainability in Latin America, Russia and Vietnam; mobilising people through dynamic citizenship and art collectives; Developing Nairobi in accordance with a Triple Top Line Approach; the emerging Chinese Eco-civilisation; and educating for sustainability in Singapore and Tasmania. The experiment reveals diversity in thought and action that crosses disciplinary boundaries, but all are united by the one specific change in the possibility of knowledge. This book takes us closer to what is intrinsic in the functioning of our world and of ourselves.

Intrinsic Capability is a portal, a point of entry

Just to be clear, it is not the kind of portal constructed to impress. It is not for example a grand entrance to a church, cathedral, temple or mosque that imposes itself on ordinary people to engender respect and reverence in anticipation of sundry revelations about the roles and influences that supernatural powers have on our lives. But it is nonetheless a point of egress to a way to radically alter lives.

Like formal religions, Intrinsic Capability changes lives by directing us to a different way of seeing the world but this new view is based on the best knowledge we now have of how the world works. Mankind has invested enormous time and resources to find out how the world functions. Consequently, we now know a great deal about the constitution of ourselves and the world we occupy. We now know for example that from the very small to the very large, nothing stands apart, alone, separately, distinct, or discrete. From the strange quantum activities of subatomic particles, to elements, molecules, compounds, cells, blood, bones, brains, people, societies, nations, continents, planets, stars, galaxies and more, we see that all is interconnected, dynamic and ceaselessly changing. Mountains, planets, stars and galaxies come and go — on a long enough time scale. Scientists now provide detailed, testable accounts of the relations that are within and are constitutive of our interconnected world and we come across them frequently in matters of health, pollution, climate change, extinctions and more. In day-to-day life, we are having to learn how best to live and develop in accordance with how the world is, how it functions intrinsically, and how it responds to our pollution and mistreatment in both the environment and our bodies. We need to live and develop institutions that work with and not against the world as we now know it — we need *intrinsic* development.

So this book is about guidance derived from science for people to live well and for developing our societies. It deals with concrete matters of building a city or managing a business as well as how best to educate and engage people in this fundamental transition. But it cannot be prescriptive for it is a book as much about "beliefs" rather than the simple, prima facie "facts" that many demand from science. This is because the word "fact" represents something far too small; in reality a little reflection will reveal the origins of any particular "fact" to be running away over far horizons. Take the "fact" of your life. You are likely to know this "fact" well enough for to be otherwise may be seen as a route to mental illness. And yet whichever way you reflect upon yourself, there is nothing straightforward to grasp, nothing to clearly and simply delineate, no constant, closed concept, no amenable story that captures your life as discrete facts that are not dependent upon and related to your family, ancestors, history, friends, society, language, assumptions, species, genetics, environment, biosphere and on to the restless, interacting compounds, molecules, atoms and subatomic particles that constitute our physical selves and which in turn have their origins in our solar system, in the remote supernovae of distant galaxies, in the big bang and so on. I am no simply "fact", and neither are you, so we necessarily handle beliefs in this book — and that requires a discerning, critical approach. Such an approach can be learned and this is one task of Intrinsic Capability.

Intrinsic Capability is necessarily a modest, reflective portal. It serves neither the establishment of enduring, fixed knowledge nor the narrow interests of power and hegemony. The best way to think of this portal is to draw from your memory, provide embellishment, and conjure in your imagination a place where you were stopped in your tracks. Where beauty, freshness, meaning, insight, hope and any other good, embracing feeling usurped your thoughts, swamped your anxieties, pummelled and pulled your rationality, took you ever-so-momentarily out of yourself, and locate there your own Intrinsic Capability portal to help access the rest of this book.

Other places where you might look to access this book include those thoughts, traditions and practices that have endured. For example, the Intrinsic Capability portal itself is nothing new:

> The best explanation I can offer is that the Shinto shrine is a visible and ever-active expression of the factual kinship — in the most literal sense of the word — which exists between individual man and the whole earth, celestial bodies and deities, whatever name they be given.
>
> When entering it, one inevitably becomes more or less conscious of that blood-relation, and the realisation of it throws into the background all feelings of anxiety, antagonism, loneliness, discouragement, as when a child comes to rest on its mother's lap.
>
> A feeling of almost palpable peace and security falls upon the visitor as he proceeds further into the holy enclosure, and to those unready for it, it comes as a shock. Epithets such as kogoshi (god-like) and kami-sabi (divinely serene) seem fully justified.
>
> — Jean Herbert, *Shinto, At the Fountain-Head of Japan*, 1967.

Indeed more of the *Intrinsic* content of this book may be found in those traditions and systems of thought that help us to better understand ourselves, our relations with each other and our relations with the natural world. Such traditions and systems can be seen as an aspect of science, as long-term experiments in living well on Planet Earth, for if they work — why reinvent them?

To summarise, Intrinsic Capability (IC) is based on science but not straightforward, hard, unreflecting scientific facts; this book is about engaging and working with facts with an open and critical mind. It is a route to a changing, evolving and ever-so wide-open thoughtscape that you can access no matter where you find yourself or whatever troubles your mind. It is as real as scientific discoveries and as potentially enduring and life-changing as a religion but it is neither created nor accessed for the one time by this book — once you recognise its knowledge and understanding, it is to be found everywhere.

Heraclitus effectively stepped through his own personalised IC portal around 2.5 millennia ago when getting his feet wet to declare that you never step into the same stream twice. More recently, the mathematician and philosopher Alfred North Whitehead provided many ways to pass through the IC portal: "There are no whole truths; all truths are half-truths. It is trying to treat them as whole truths that plays the devil." (Whitehead, 2001 [1954]).

The IC portal is a way to explore and revise freedom. Take "Free Markets" for example, the concept is widely accepted and applied with such a measure of certainty and closure, that it is adversely, perhaps irreversible, damaging the planet and its inhabitants. We urgently need to free ourselves of free-markets.

Similarly in microeconomics, the formal, detailed financial accounts provided of the activities of businesses engender stagnation. The Jesuit Order faced and overcame this problem centuries ago with the use of a financial accounting system based on a cash box requiring two different keys held by two different people. As Quattrone (2015, pp. 23–24) explains:

> The existence of a material padlock operated by two keys allowed a continuous performance around the notion of rational and legitimated behaviour when cashing and spending money. This continuous mediation around and questioning of the rationale informing the opening of the box made the Jesuit rationality an unfolding one.

> As with every graphical representation, the translation of cash movements into accounting inscriptions would have reduced the multifaceted nature of the Jesuit administration to a mere financial matter, leading to an incontrovertible financial theoria and to unreflective actions. This unreflective representation would have deprived the Jesuit member of that indifference that characterised the Jesuit's self and made him able to exert wise judgment. Jesuits' representations, whether dealing with spiritual or financial matters, were meant to manage and not reduce the 'virtuality of the possible' of Jesuit behaviours (Barthes, 1971). They were not aimed at producing matters of fact but matters of concern.

The IC portal does not have two keys since it is an open door: instead it has two sides. On the one side, there is a set of facts, hard, indisputable and accepted by many people without reflection. For convenience, we can label this side *Modern*. The IC meaning of *Modern,* its thoughtscape, is one aspect of the subject of this book to be explored thoroughly. But to start the exploration, as a first guide, key *Modern* features that have our attention include its quirky use of Rationality which is instantiated in its understanding of Capitalism, economic exchange, tradition, development, rewards and a limited lifetime of achievement.

When stepping through the IC portal, we follow the footsteps of two colleagues now deceased, Grey Beard and the Older Boy (Birkin and Polesie, 2012). These two characters recorded their encounters over several decades as they passed through the *Modern* thoughtscape and its derivations. They described the IC portal as an example of an episteme change after Foucault (2002 [1966]).

They passed through the portal to describe a new age, an emerging episteme, which they named *Primal*. They were both accountants so their view of the *Primal* episteme included ideas for economics and business which include a transition from *economic maximisation* to *ecological optimisation*, a Triple Top Line business appraisal method and a conceptual shift from "bulldozer" to "dinghy" companies. They also discovered the *Primal* thoughtscape to be rich in ecological relations and a wealth of time-tested traditions available to guide cultures and lives.

When they talked of their experiences and understanding, they found many others who were thinking the same way. In time, it became apparent to Grey Beard and the Older Boy that large number of people from all quarters were either in transition, at the cusp of changing, or had fully changed, were already truly *Primal*. The pair slowly came to recognise that the *Primal* was in a sense ever thus; ever going back to basics, changing, revising, refreshing, restating and reapplying itself. They talked, debated, sweated and struggled and came to an obvious conclusion known since the beginning of time to all that participate from philosophers to worms, from seagulls to neutrinos, that the only constant is change. Change is the way of the world.

The pair explained their observations of the epistemic transition as a movement:

From Modern abstract rational belief systems (such as neo-classical economic theory and mainstream financial accounting).

To Primal pragmatic empiricism (based on the findings of science supported by ancient traditions that have endured).

They also noted that the *Modern* abstract rational belief systems were very good at providing clear factual explanations of how to organise our institutions and our lives and how to get more stuff. It is the promise of more stuff that motivates many people, sometimes excessively enough to be reminiscent of the feeding frenzies of sharks. ... but it is far, far more dangerous.

Although very old, Grey Beard and the Older Boy thought they had better assist the transition and help establish the emerging new age, the *Primal*. To this end they created an online presence, Intrinsic Earth to help people to focus and mobilize "... a social movement exploring new possibilities for knowledge, connecting healthy & vibrant human communities and supporting people, governments and businesses in the transition to a sustainable world." (Intrinsic Earth, 2018).

They also got a publisher to promise to publish a book in which people in the transition would write about their own thoughts and observations. Eventually, posthumously, the book was published and it is now in your hands. The book's content is very varied as it represents the reach and depth of the transition in many quarters:

- Working for better lives for all;
- What we have known for thousands of years and what we have just discovered;
- Changing our understanding of success in business by changing the practices of marketing and accounting;
- Building cities for people and not capital;
- Educating to challenge beliefs and assumptions and moving forward to think about how things could be otherwise;

- Engaging, empowering and mobilising society;
- An emerging civilisation;
- Art, artistic collectives and sustainability.

It is a range of topics that taps into the diversity of the world itself. Everyone they asked was keen to contribute. They all saw in the book a unity and collaboration that strengthened and enlarged their own arguments and struggles. What now follows are summaries of their contributions as they presented in this book.

Chapter 2: Traditions

Strategies for epistemic survival: Slave and scholar rebels in the boundaries of Hispanic America

Alejandro and Mónica are personally aware of the range of other than Modern knowledge and capabilities in South America. For them, these alternatives provide so many other options for development especially those that link ecological knowledge, people, and alternative ways of living. They see these alternatives in an ontological sense as different ways of being or existing. However the very presence of these alternatives interacts with Modernity to create tensions, the wrong sense of urgency and sometimes sacrifices are required.

A Russian perspective on the limitation of existing account reporting practices for business development and management

Tatiana and Olga delve into Russian tradition to better understand the Modern Russian focus on business and accountability reporting. Their study is from the spread of the Enlightenment into Russia, to the Soviet state, the collapse of the USSR, perestroika and the impact of sustainability on Russian companies. Within this grand history, Tatiana and Olga seek to re-establish the humans side for the "thinking behind Russian business development and reporting has moved away from the intimate and tender understating of the relations between the humans and nature".

Their arguments draw on Russian proverbs and the great Russian novelists. They leave us with a tantalising question about the emergence of a new Russia.

Chapter 3: Ecocentric business and marketing: Ideology, reality and vision

This international and multidisciplinary team tackles an emerging tradition, culture, or worldview that provides a very different way of doing business. This change is found to be arising from deep within the modern business world, and is represents a profound challenge to the status quo. By examining some basic assumptions in modern business and by identifying the adaptions and evolutions now underway in the name of sustainability, Helen, Adam, Véronique and Joëlle reveal how businesses are coping with all the relations to the world that always have been there, but have been overlooked in modernity. From a foundation in scientific and systems studies, the authors use an ecological perspective to challenge to consumerism and its principal agent, modern marketing. The authors skilfully describe the changes as a shift from an anthropocentric to ecocentric ontology, whilst providing corporate leaders with realistic and operational transformation strategies, redesigned business models and opportunities in new products and services.

Chapter 4: Towards an ecological accounting

People experience and make sense of the world... and they give accounts. Accounts take many forms from art and music to verbal traditions, literature, diaries, and company performance reports. We are an account-giving and account-receiving species. Such accounts help us to navigate and to make sense of our world, but such accounts can also be malign and blind us to the destruction of ourselves and of the world we inhabit. In this chapter, Rob explores the extent to which the social and environmental accounting might help in closing the gap between (i) the formal organisational accounts of Modernity that ignore society and ecology relations and (ii) the possibilities of a humankind with a more sensible relationship with its own and with ecological nature.

Chapter 5: Ecological civilisation

China's emerging ecological civilisation

In this contribution, Liu and John examine China's emerging ecological civilisation. They consider that China's philosophical traditions, based on Buddhism, Daoism and Confucianism, have created an embedded ecological wisdom that is still highly influential in China today. These philosophies and accompanying traditions are argued to underpin Chinas emerging ecological civilisation which they explain an example of episteme change. In addition, the changes in China can be linked to the Capability Approach of Amartya Sen in so far as the Chinese people are regaining basic capabilities of which they were previously denied: to be able to breathe clean air; drink pure water; to work with fair working conditions; and to move and work wherever they wish.

Can Vietnamese traditional values drive 'Stainability'? an Emic perspective

Here Lien tries something she has never done before and that is to write an emic account of Vietnamese culture and its impact on their relationship with Nature. An emic perspective relates to the perspective of a person living within a culture so it is rich with insight and understanding about their own way of living. This contribution is enriched by drawing on research by other Vietnamese and foreign authors.

Chapter 6: African development and management

Development in Nairobi: Three into One does not GO!

In Nairobi, over the past decade, low-density suburban districts have experienced rapid vertical densification resulting in a preponderance of high-rise apartment blocks where single-dwelling units historically held sway. The emergent new apartments are "modern" in the universal sense of the word representing the architectural expression of global capital.

Situated in the global South, the buildings differ very little from architectural expression in the global North and bear little relationship to local and traditional modes of architectural expression. Yet, an opportunity exists to celebrate local culture and use it for locally appropriate modes of architectural expression. In addition, the 2016 *Ngorongoro Declaration* recognised that culture is a crucial ingredient in the path towards achieving sustainable development. In this chapter, Collins examines the tensions in the development of Nairobi city between, on the one hand, local cultures and traditions and, on the other, globalised capital.

A Tale of two theoretical cultures: Expanding the application of Habermas' communicative action and Post-colonial theories in management studies and practices in Africa

Post-colonial theory has undoubtedly earned itself a deserving stature but it has achieved this with few contributions from African academics practicing in Africa. In this chapter, Sharif and Adeyinka help to redress this imbalance. It considers that post-colonial theories offer some close similarities with some continental theories including such as Habermas' Theory of Communicative Action. The Theory of Communicative Action with its ethos of mutual communication aimed at achieving an "ideal-speech" situation exude some embracing tenets of post-colonialism. An illustrative case study is provided in this chapter using Ghana's mining industry. A theoretical understanding is useful in achieving a degree of critical relevance. In essence, an application of combined Habermas' Theory of Communicative Action and post-colonial theory showcases how the existential relationship with Africa and the multinational world of mining emanates issues of accountability and transparency.

Chapter 7: Educating for sustainability

In this chapter, Chris and Kim review the history of sustainability education. But sustainability education, like sustainability itself, remains a contested space in which to think and work. There is however a framework that has opened space for society to foster knowledge-building oriented schools, universities and organisations without walls to study and practice our responsibilities *on sustainability*. Even so, it is in the nature of sustainability that it brings into questions fundamental beliefs and assumptions about the world and our positioning as humans within it. Consequently, Chris and Kim conclude that education for sustainability is primarily about challenging these beliefs and assumptions and moving forward to think about how things could be otherwise.

Chapter 8: Emerging business values

Changing Values

Ecus Ltd. are a multidisciplinary environmental consultancy with offices across the UK. Nick is contributing to the Intrinsic Capability project to help identify the inherent connections people have to the environment and how these can be used by organisations and business to better understand and face their customers.

This contribution on "Changing Values" aims to discuss the response of organisations and business to the intrinsic value of the environment to their customers as well as highlighting opportunities for collaborative working in allied organisations and innovation in business practice.

Social Entrepreneurship in the Agrifood Sector: Smallholder Farmer Co-operatives

This contribution will focus on providing a narrative of the business model developed by the Centre for Sustainable Agriculture (CSA) in India, based on the principles of social entrepreneurship. The CSA has developed this business model by bringing together over 50,000 smallholder farmers to create farmers cooperatives, and a federation of these cooperatives in the form of a producer company that has created a chain of retail stores to sell the farmers produce directly to the end consumer. The CSA was developed in response to the challenges currently being faced by smallholder farmers in India in being able to achieve sustainable livelihoods. The inability of smallholder farmers to achieve and maintain sustainable livelihoods is due in part to what is known as the agrarian crisis.

Chapter 9: Mobilising citizens

Dynamic Citizenship: Participatory democracy on local level

This chapter describes a number of projects in Sweden and Bosnia-Herzegovina where the concept of dynamic citizenship has served as a model for democratic development. The reinforcement of the democratic process have been the overall perspective of the projects, where the relationship between the citizen and Dynamic Citizenship consists of three "democratic qualities" — the degree of participation, the degree of influence and the degree of involvement. These three "democratic qualities" can only be reached if the terms of openness and insight, political equality and meaningful participation have been clearly defined as a political ambition. In conclusion, Per-Eric argues that *political equality* refers to everyone's right to participate, influence and be involved in social development regardless of ethnic or cultural background, social or economic status, sex or age.

Engaging Imaginaries: The Role of Artistic Collectives for Transdisciplinary Sustainability

Sustainability requires the integration of different forms of knowledge, not only through the combination of interdisciplinary collaborations, but also through collaborations that transgress the borders of academia — transdisciplinary forms of knowledge. The investigation of environmental aesthetics and arts-based research provides a fruitful platform for such transdisciplinary exchanges. In this chapter, Andressa presents an example of the work of the Brazilian artistic collective "Coletivo Líquida Ação" which focuses on the topic of water: access to water, water use, abuse, pollution, draughts and the complex imaginaries that surround such topic and give meaning to this substance. The focus is the project "Foz Afora" realised by the collective between June and October 2017. This project highlights the current reality of several communities affected by the collapse of a mining-tailings dam and the devastation of the Doce River that occurred in Brazil in November 2015. The aim of this chapter is to investigate, through this example, the multiple aspects of environmental aesthetics and the role of artistic collectives in engaging with local imaginaries towards nature and the environment and in generating or improving environmental awareness and meaningful actions for sustainability.

Chapter 10: A reflection by Frank Birkin and Thomas Polesie

This Intrinsic Capability portal is open for all — enjoy!

Chapter 2

TRADITIONS

2A — Strategies for Epistemic Survival: Slave and Scholar Rebels in the Boundaries of Hispanic America

Alejandro Balanzo[*,‡] and Mónica Ramos-Mejía[†,§]

*Universidad Externado de Colombia, Bogota, Colombia
†Pontificia Universidad Javeriana, Bogotá, Colombia
‡a.balanzoguzman@utwente.nl
§ramosm.monica@javeriana.edu.co

> *The epistemic diversity of the world is potentially infinite.*
> *There is no ignorance or knowledge in general.*
> *All ignorance is ignorant of a certain knowledge, and all knowledge*
> *is the overcoming of a particular ignorance.*
> *There are no complete knowledges.*
>
> (Santos *et al.*, 2007)

2A.1 Small towns in the boundaries

It was the early 17th century when the Spanish crown decided to grant freedom to San Basilio de Palenque. The decision, a royal chart by

Antonio Ortiz de Otálora, followed a report from Baltazar de la Fuente to Antonio de Arguelles. This report related the state of affairs of the warfare against the rebel slaves that had escaped and built a little town in the mountains at the south of Cartagena de Indias. This is the first free town of America, a little dot in the map that has been there ever since, adding to the multicultural landscape of Colombia (Arrázola). With less than 1% of the total afro-descendent population, *palenquero* is one of the four possible afro-descendent Colombian identity references (DANE, S.F.).

Yet, San Basilio has never ceased to exist in the boundaries. Even today, more than 200 years after Western-minded creoles defeated the Spanish colonial power chasing the dream of an independent Republic. Expectedly, Palenque San Basilio is but one of many human groups struggling to survive in the shadow of the Western. Many other afro-descendent groups settled in Pacific Colombia, while a large number of small indigenous communities kept little territories all over the country. While Spanish is still the official Colombian language, it is known that the total diversity amounts to 63 different languages (DANE, S.F.).

San Basilio de Palenque has a creole language, mixing African and Portuguese roots on a Spanish base. This culture has a deep sense of musicality, expressed to deal with and translate every social aspect through a rich rhythmic base. They also have specific beliefs and rituals to walk through the paths of life and death, and a very particular social structure, based on *kuagros* — age groups — rather than biological families. Last, they have developed a number of natural medicinal and agricultural technics (De Friedemann, 1987).

This richness was bound to disappear almost entirely by the 1960s, as an effect of various external pressures. Racism, exclusion and being invisible were common as the 19th century advanced. If not hiding their identity to prevent ill treatment from others, white and black alike, *palenqueros* would be pushed by the government to work on private sugar cane fields. They would leave behind not just their lands but also their customs, as the *palenquero* song sings (Tabala, S.F.).

It was culture valorisation, starting in the early 1970s, what prevented this culture spiriting away. As a teaching initiative, based on the educational method of asking the elders, a number of teachers managed to create an ethno-educational strategy. With time, this strategy would be the bedrock of a prominent role of cultural valorisation through musical festivals, academic workshops and policy incidence. Armed with a set of organisations giving it legal status to engage in varied quests, *palenqueros* managed through the years to create a regional state programme on ethnic education, as well as to achieve UNESCO's recognition as immaterial patrimony in the late 1990s, thus ensuring to setting in motion of a number of measures for the protection of their culture (Soto *et al.*, 2009).

They were not fighting alone. Also by the 1970s, afro-descendants from Pacific Colombia mobilised, pushing the government for regulations regarding collective land ownership. They managed to pass Law 70, creating a platform ensuring means to preserve afro-descendants right to land. Although Law 70 did not aim at creating territorial public administrative offices, afro-descendants took it as a chance to create local bodies with sufficient steering capacity (CEPAC, 2003). In little time San Basilio de Palenque would institute *Ma Kankamana*, its own council, as a local authority allowing deliberation, economic planning and interlocution with territorial administrative offices. In some other places they also set for them functionality as cooperatives (Balanzo, 2016).

2A.2 Co-creating hybrid worlds

The short story of a small town with great cultural value struggling to survive amidst external pressures is a common story in Latin America. A story that can be read in the light of a quest of human freedom aiming at deploying meaningful autonomies (Sen, 1999). But it is also a story that can be read in the light of resistance: the various games being played at the boundaries so as to bargain better positions (Balanzo, 2016). Those games, here described, show moments of direct opposition, tentative compromise or collaboration at the boundaries. And because of that, they add to the rather tense landscape of social construction of Latin American institutions.

Interestingly, a stream of Latin American thought describes a similar picture: there has also been intellectual resistance and a somewhat different form of what Gieryn (1983) called boundary work. These approaches stem from the Latin American discussion about existence in the boundaries — being *the Other* — and epistemic diversity.

This discussion arises from a clear awareness of the contradictions of Western modernity. Witnessed from afar, the ruthless epistemic violence of modernity is strikingly evident (Lander and Castro-Gómez, 2011). The modern project dreamed of objective science, universal moral and self-regulated laws and arts. This project, it is argued, nurtured the kind of European self-consciousness by means of which the colonial mindset has been so far justified: a modern (Christianly blessed, wealthy, advanced, civilised, developed, first), and the others (soul-less, savages, poor, lacking, underdeveloped, third). Montaigne saw very well the contradiction: "we can call them barbarians regarding our reasoning rules, but not regarding ourselves, who outdo them in every kind of barbarism".

Although every culture is ethnocentric, argues Dussel, modern European ethnocentrism has been the only one attempting to identify itself with universality-globality. Modern Euro-centrism confused abstract universality with concrete globality while building Europe as a hegemonic centre (Dussel, 2000). There is little irony in the fact that this contradiction was inherited by Latin American republics. Although local elites saw European contradiction, their motives for rebellion were geo-political. They could have not been epistemic, for they shared — and often longed for — the Western mindset (Mignolo, 2000).

Latin American thought includes critical — sometimes radical, but also often creative and compromising — views on the post-colonial present, but also insightful views on Western modernity and the rationales and roles of scientific expertise. Naturally, the critique starts by pointing out to how it is that the cannon of social sciences themselves most be deconstructed in order to counteract its role as a legitimation device of the modern monoculture regime. The bottom line here points to the fact that there is no global justice without cognitive justice, and therefore the monoculture of scientific knowledge should be replaced by an ecology of knowledges (Santos *et al.*, 2007).

This position draws on various sources. Some of them, such as the feminist critique, Indian post-colonial thought and African thought, come from foreign scholar traditions. Some others link to Latin American emancipatory thought, including, for example, the pedagogy of the oppressed (Freire, 2000) and participatory action-research (Fals-Borda and Rahman, 1991). This stance aims at challenging the common assumption of European history as universal history, as well as the common assumption of modern thought and institutions as universal thought and institutions (Lander and Castro-Gómez, 2011).

According to Montero (1998), this leads to the possibility of a Latin American capacity to observe and being while standing on *Another* perspective. A perspective placed on the *We*. This perspective, Montero argues, is built on:

- Particular notions of community, participation and popular knowledges as constitutive building blocks as well as relational *epistemes*.
- The notion of liberation through praxis, mobilising consciousness and critical senses in order to denaturalise the canons of absorbing-building-being in the world.
- A re-definition of the role of the researcher, reckoning of the Other as Self, and therefore the redefinition of the subject–object dichotomy in favour of a role as a social actor and a knowledge re-creator.
- The historical, undetermined, undefined, unfinished and relative character of knowledges.
- The multiplicity of voices, life worlds, and epistemic diversity.
- The perspective of dependence and, afterwards, resistance.
- And the permanent tension between majority and minorities and the alternative ways of doing-knowing".

2A.3 As means of a conclusion

The story of a small town in the boundaries plus one stream of local scholar work is still the story of a small town in the boundaries. Various repertoires have been set in motion in the case of San Basilio de Palenque: scholars have helped offering texts which communities often use as tools for making visible and to sharpen their discourses, e.g. (De Friedemann, 1987; Soto *et al.*, 2009). They have also suggested hybrid solutions, both at community and policy level, to achieve coexistence of identity preservation and capitalism, e.g. Schejtman and Ranaboldo (2009). Very often the Latin American scholar finds herself or himself at the cross-roads of the double boundary: on the one hand, serving "universal" modern science while protecting local epistemic diversity; on the other, building local knowledge references while collaborating to disrupt, create or reflect on social institutions themselves.

Overall, Latin American social and institutional landscape shows the marks of hybrid compromises amidst inescapable boundaries: trails of modern, indigenous and afro-descendant coexist in overlapping, and often contradicting, regimes. Its tension built on the struggle of ruthless exclusion; the fight of social movements or the creative co-construction of memory, markets, networks and law.

For the last four decades *palenqueros* have been relatively successful in manoeuvring within the developmental discourse, using their "cultural space" as their main bastion. However, besides interesting anthropological findings, little have social scientist learned from their social richness. We believe that by recognising a change of epistem, researchers would be able to find exchanges and relations that go beyond the economic dimension, in order to explore *palenqueros'* understandings and practices related to Nature and, therefore, bringing traditional knowledge into sustainable development science.

2B — A Russian Perspective on the Limitation of Existing Account Reporting Practices for Business Development and Management

Tatiana Dyukina[*,‡] and Olga Cam[†,§]

*Saint-Petersburg University, St. Petersburg 199034, Russia
†University of Sheffield, Sheffield S10 2TN, UK
‡dtodom@mail.ru
§o.cam@sheffield.ac.uk

The origins of modern reporting in Russia trace back to 1721 during The Emperor's All-Russian Peter I (Peter the Great) ruling period. As the head of state Peter I led a reformation campaign aiming to modernise Russia's social and political systems including the development of reporting systems. Inspired by Enlightenment ideas from the West, the Emperor promoted European law-enforcement practice and reformation of the education system. This also included the creation of the first business school. Such changes had a powerful impact on the development of early Russian reporting systems.

Following the Soviet revolution, the development of reporting during the post-revolutionary period and years of the Soviet economy changed its course. With the support of the revolutionary views along with ideas from national culture and traditions, a new centralised regulation system of formal institutions for reporting was developed. The Soviet period reporting system adhered to the principles of domination of public property as a means of production, which resulted in the creation of a uniform order of documentary registration.

The new era of accounting development in Russia started in the late 1980s during the perestroika time, which followed by the collapse of the USSR. Accounting for the activities of both commercial, and non-profit organisations, as well as foreign economic activity during reorganisation, demanded a new reporting paradigm. One which could reflect the new economic reality. Thus, since the beginning of the 1990s reporting reform was set in motion, resulting in the development of Russian national standards of accounting and reporting (Accounting Regulations) and a gradual rapprochement of Russian standards with the International Financial Reporting Standards (IFRS).

While much attention was given to the redevelopment of financial reporting, the first non-financial reports appeared in Russia around 2000. Preparation of non-financial reporting by the organisations in Russia is voluntary. Currently, non-financial reporting mainly attracts the interest of larger Russian companies who see their business purposes as closely connected with corporate social responsibility. However, overall the number of companies providing non-financial reports is growing. Today there are Russian companies which use the concept of sustainable development to define the purposes and problems of financial and economic activities. The actions of these businesses are guided by the identification and analysis of the actual and potential economic, ecological and social influences. As a result, there is some increase in the transparency of companies' reporting. The Russian Union of Industrialists and Entrepreneurs which runs the National Register of companies publishing non-financial reports indicates that since 2000 around 254 out of 764 filed reports were in the field of sustainable development. This is made up of 70 ecological reports, 318 social reports, 122 integrated reports and 25 industry reports (RSPP, 2018). While the presence of non-financial reporting is a positive indicator of a growing understanding of the need to consider and reflect upon the wider impact of business actions, the numbers of reports submitted are still a very small proportion of the number of existing organisations. In research on Russian reporting practices, Rognova (2013) suggests that some reporting problems lie within the tools used for business creation and development. She points that in Russia these challenges and restrictions stem from the issues of trust, prospects and reliability. She notes, however, that these problems are not unique to Russia but of a universal character.

Given the significant social and political change in the Russian Federation since the collapse of the USSR and the overall novelty of accounting in a post-planned economy era, relatively little attention has been given so far to the topic of accounting and ecology or accounting and sustainable development in Russian scholarship. Nevertheless, there is evidence of narrow measurement and limited consideration given to the use of natural resources in the accounting practices adopted in the post-Soviet era. Murueva's (2007) study of the ecological aspect of accounting practices with a focus on Russian forest industry explores the limitations of currently adopted accounting practices in ascertaining "ecological" expenditure. The author concludes that current methods of cost allocation have many faults and are not suitable for the purposes. Instead, Murueva argues for the need to develop a complex methodology when planning for the use of natural resources.

It appears that in the race for modernisation and growth the thinking behind Russian business development and reporting has moved away from the intimate and tender understating of the relations between the humans and nature. For many centuries the knowledge and wisdom of these relationships were passed across generations through ancient idioms and proverbs. These traditional insights passed from generation to generation not only survived the test of time but are a big part of Russian culture. Such ideas offer time-tested foundations for rethinking business development and reporting practices.

The Russian inter-generational sagacity urges humankind to avoid short-term views in life; to consider both the power and the fragility of nature and to observe and respect the dependence of humankind on nature's way. For example, a famous proverb:

> *Бережливость хороша, да скупость страшна.* It translates as "Economy is a good servant but a bad master" and refers to the underlying assumptions in business planning and performance assessment. The proverb questions the prevailing cost-cutting approach as a means to optimise business.

Traditional living in Russia depended on natural resources. Because of this, many proverbs advice on approaching nature with thought and care when doing and/or planning to do business:

Срубить дерево — пять минут, вырастить — сто лет (To chop a tree — five minutes, to grow it — one hundred years);

Кто земле дает, тому земля втройне отдает (To those who give to Earth, will reap the benefits her generosity).

To the current generations, these intrinsic ideas offer guidance to designing and building a business. The wisdom urges humanity to see themselves as part of natural arrangements and suggests for humans to re-think their position in nature to allow for the sustainable development of Earth.

Deeply embedded in the Russian folklore concept of Mother Nature and the need for humanity to understand its place in nature also echoes in the thinking of many Russian philosophers and writers. Thus already in 1861 in his "Fathers and Sons", Turgenev (1971) pointed out that the people started to forget where they came from and that nature is their first home and that nature that created the man. The author is anxious that despite such overwhelming arguments, everyone fails to pay sufficient attention to the environment despite that all our immediate efforts should be dedicated to saving it!

For Dostoyevsky (1994, p. 253):

Соприкосновение с природой есть самое последнее слово всякого прогресса, науки, рассудка, здравого смысла, вкуса и отличной манеры.

The idea is translated as follows:

"Becoming one with nature it is the last word of any progress, science, common sense, good taste and excellent manners."

In Tolstoy's writings the reader will see the condemnation of a society where human placed above nature. The author (Tolstoy, 1985, p. 54) wrote in his diary:

В безнравственном обществе все изобретения, увеличивающие власть человека над природою, — не только не блага, но несомненное и очев идное зло

In English translation the passage can be read as:

"In an immoral society all inventions that aim at increasing the power of humankind over nature — are not only not benefits, but undoubted and evident evil".

Such rich heritage in understanding nature and human coexistence offers fruitful soil for new business thinking. Guy Eames, a co-founder of the Russian Green Building Council, notes that a New Russia is emerging. Built on the strong environmental awareness raised in the Soviet times the New Russia is not only absorbing the international sustainability concepts but also turns to its traditional ideas to develop new technologies for a sustainable living.

Already some Russian grassroots businesses are coming through whose technological process build not against, but from the needs and offers of nature. Novie Biotechnologii (*ООО Новые Биотехнологии*) is an example of a company where sustainable development approaches embedded as a founding principle. With its project — Zooprotein the company offers organic feed and fertilisers from recycled organic waste. The organisation prides itself in that the technological processes, which use modern scientific research, grounded need to rethink the impact of the growing human population on the fragile planet Earth (Zooprotein.com, 2017). Novie Biotechnologii is just one of many new "dingy" companies (Birkin and Polesie, 2012), who with the right support, can learn to navigate the open ocean of Mother Nature.

Chapter 3

ECOCENTRIC BUSINESS AND MARKETING

Sustainability, Epistemology, Ecocentric Business and Marketing Strategy: Ideology, Reality and Vision

Helen Borland[*,¶], Adam Lindgreen[†,‖], Véronique Ambrosini[‡,**]
and Joëlle Vanhamme[§,††]

*Aston Business School, Birmingham B4 7EQ, UK
†Copenhagen Business School, 2000 Frederiksberg, Denmark and
University of Pretoria's Gordon Institute of Business Science,
Johannesburg, South Africa
‡Monash Business School, Caulfield East VIC 3145, Australia
§EDHEC Business School, 59057 Roubaix Cedex 1, France
¶h.m.borland@aston.ac.uk
‖adli.marktg@cbs.dk
**v.ambrosini@monash.edu
††Joelle.vanhamme@edhec.edu

3.1 Introduction

Sustainability and marketing make somewhat unusual bedfellows in intellectual discourse (Banerjee *et al.*, 2009) in that they traditionally take opposite sides on the consumption continuum (Menon and Menon, 1997). Yet, sustainability is advancing rapidly as a viable ideology in political, economic, technological and academic circles, even though little theoretical, empirical or strategic research has attempted to understand it in depth (Bansal and Roth, 2000; Kilbourne *et al.*, 2002; Sharma *et al.*, 2010). Understanding the strategic nature of sustainability and marketing theory development is even less well documented (Borland, 2009a; Kilbourne, 1998; Sharma *et al.*, 2007; Varadarajan, 2010).

Part of the issue may be that sustainability is rooted in several scientific disciplines and does not belong to any one. It is quintessentially interdisciplinary and represented by varied theories and laws, including systems theory, ecosystems theory, the laws of thermodynamics and Gaia theory (Borland, 2009a). Collectively, these theories and laws explain the behaviour, homeostatic balance and maintenance of life on Earth (Borland, 2009a; Lovelock, 2000). However, the UN's World Commission on Environment and Development (1987) prompted widespread adoption of an anthropocentric view of sustainability (Borland, 2009a; Purser *et al.*, 1995; Sharma *et al.*, 2007). This view prioritises a human bias and has generated sub-disciplines such as environmental management, sustainable development and environmental resource management (Porritt, 2007; Purser *et al.*, 1995), which put human needs and wants or further human expansion and development above the survival and development needs of other species. To delineate the properties of sustainability, Belz and Peattie (2009) instead suggest a framework that features a holistic and systems-based view, an open-ended timeframe, a global perspective that focuses on ecological sustainability rather than economic efficiency and recognition of the intrinsic value of nature. Yet, sustainability also demands recognition of the finite limits of the nature as a source of resources and a repository of waste, and it distinguishes between (impossible) unlimited economic growth and sustainable growth, which implies a qualitative improvement in means and ends (Ekins, 2000; Guest, 2010) through improved health and well-being for all species. Because ecological sustainability implies a fundamentally different way of looking at the world, as well as marketing's place in it, it demands an expansion of the limits of marketing enquiry (Grönroos, 2007; Hult, 2011; Varey, 2010). That is, the marketing discipline must adopt a more macro-focus and more multidisciplinary methods (Cronin *et al.*, 2011; Kilbourne, 1998; Mittelstaedt and Kilbourne, 2006; Varey, 2011).

The modern marketing philosophy instead emphasises greater consumption as a societal end-point (Schaefer and Crane, 2005), perpetuates an anthropocentric ideology and aims to maximise corporate profits by satisfying the preferences and choices of individual consumer targets (Ellis *et al.*, 2011). Such a view produces conventional, cradle-to-grave products and services that firms label "green" or "eco", misleading consumers into thinking they are helping the environment (Peattie, 1999). Curry (2011) delineates between "light green", "mid green" and "deep green" products and services. But perpetuating an anthropocentric ideology through conventional marketing activity cannot lead to sustainability (Hart and Milstein, 1999), especially in the face of exponential global population growth, resource depletion, over consumption, waste accumulation and habitat destruction. Addressing such manifestations of the excesses of human activity through existing mental filters and mindsets will have little effect on future prospects (Borland, 2009b; Bosselmann, 1995). In this chapter, we therefore consider a different approach to strategic marketing that is based on ecological sustainability and ecocentric epistemology. Can strategic marketing truly be reconceptualised on ecocentrism and ecological sustainability?

Ecological sustainability is defined as the capacity for continuance into the long-term future, by living within the constraints and limits of the biophysical world (Porritt, 2007). It represents a goal, endpoint or desired destination for the human species as much as for any other species and can be explained, defined and measured scientifically. Sustainable development instead refers to the process for moving towards sustainability; it implies trying to achieve sustainability but often seems poorly defined and difficult to measure. To achieve a sustainable human future, sustainable development generally includes social and economic elements, as well as environmental ones, though Porritt (2007) considers those elements secondary goals, because all else is conditional on living sustainably within the Earth's systems and limits. The pursuit of ecological sustainability thus is non-negotiable (Porritt, 2007; see also Mort, 2010).

The urgency of the ecological sustainability predicament drives the search for new ways of living and conducting business (Mort, 2010). Yet, many corporate initiatives towards what the firms perceive to be sustainability are simply efficiency drives or competitive moves (Unruh and Ettenson, 2010) — falling far short of actual strategies for ecological sustainability. To suggest true ecological sustainability strategies, we adopt an interdisciplinary, or transdisciplinary (Gladwin *et al.*, 1995), approach to discern what business and marketing strategies might look like if they were underpinned by environmental and ecological science. In particular, an ecocentric epistemology offers an alternative cultural and mental framework that focuses on the whole system or ecosystem and the balance of all species and elements (i.e. rocks, water and gases of the atmosphere). Humans thus move from their cosmologically central and egocentric position, in which the whole of nature exists only for their exploitation with no intrinsic value (Gladwin *et al.*, 1995; Kilbourne, 1998; Kilbourne *et al.*, 2002; Purser *et al.*, 1995), to a more balanced site in the larger system that demands greater appreciation of and respect for other species and planetary resources (Borland, 2009a; Du Nann Winter and Koger, 2004; Porritt, 2007; Shrivastava, 1995).

To this end, we take an ideological and visionary perspective that is grounded in the reality of our modern lack of ecological sustainability, which currently creates the prospect of ecological collapse and the loss of some or all resources and eco-support services on which we depend. Therefore, we begin by taking a step back to investigate the epistemological basis of our knowledge about the world. Section 3.2 then examines how strategy develops differently according to the two different epistemological ideologies presented. This leads us to consider the introduction of strategies for ecological sustainability based on these different epistemological assumptions, namely transitional and transformational strategies. The chapter then outlines what an ecocentric strategic marketing vision would look like and includes practical steps and examples of this. Finally, we consider the general applicability of ecocentric business and marketing strategies and provide six universal foundational premises, as part of future conceptual and theoretical development. Managerial implications and thoughts for further research are also discussed.

3.2 Anthropocentric and ecocentric epistemology

3.2.1 *Anthropocentric ideology*

An anthropocentric ideology embraces the notion of human exemption: Unlike other species, humans are exempt from the constraints of nature, and the whole of nature exists primarily for human use with no inherent value of its own. This notion is reflected in widespread beliefs about the benefits of abundance and progress, pursuit of unlimited growth and prosperity, faith in science and technology, and commitments to a laissez-faire economy, limited government planning or intervention and private property rights. This modern Western worldview posits that land not used for economic gain is wasted and that people have the right to develop land and do with it as they see fit (Kilbourne, 1998; Purser *et al.*, 1995).

Purser *et al.* (1995) propose limits to anthropocentrism, including primarily that it offers no overall survival plan (see also Kilbourne, 1998). A consumption rhetoric, also termed helpfully the "social logic of consumerism" by Smart (2010), is a means to an end that lacks an endpoint, with no overall goal — human or otherwise. Economic growth in this ideology continues unlimited and unchecked, until complete destruction of the physical environment and natural resource base occurs (Diamond, 2006).

Yet, the anthropocentric ideology is well embedded in human society, likely because it helps those who benefit most maintain their power and wealth. The conceptual differentiation of the socially constructed hierarchy dictated by the anthropocentric human–nature dualism allows people to construe nature as unlike them, which offers support for the claim that humans are morally superior to non-humans and thus justified in dominating nature (Purser *et al.*, 1995). Such an anthropocentric attitude essentially denies any inherent worth to nature (Gladwin *et al.*, 1995), as we depict in Table 3.1.

Table 3.1.   Summary of Ecocentric Strategy Development

Philosophical Component	Transitional Strategies	Transformational Strategies	Selected Sources (Chronological)
Epistemology	Anthropocentric perspective	Ecocentric perspective	Leopold (1970), Rolston (1994), Naess (1995), Purser et al. (1995), Gladwin et al. (1995), Diesendorf and Hamilton (1997), Kilbourne (1998), Dunlap et al. (2000) and Schultze and Stabell (2004).
Paradigm	Dominant social paradigm	Ecocentric responsibility paradigm or new ecological paradigm	Purser et al. (1995), Gladwin et al. (1995), Kilbourne (1998), Dunlap et al. (2000), Lovelock (2000), Kilbourne et al. (2002), Banerjee (2002), Schultze and Stabell (2004), Banerjee et al. (2009) and Borland (2009a).
Value set	Rational Instrumental Egocentric Exemptionalist Narcissistic Economic rationality	Emotional Intrinsic, value-driven Spiritually advanced Ecolibrium Empathetic Ecological rationality	Rolston (1994), Shrivastava (1995), Naess (1995), Purser et al. (1995), Gladwin et al. (1995), Bosselmann (1995), Kilbourne (1998), Zohar and Marshall (2004), Du Nann Winter and Koger (2004), Stead and Stead (2004), Porritt (2007), Ketola (2008), Borland (2009a) and Linnenluecke and Griffiths (2010).
Scientific approach	Reductionist Deconstructionist Empirical	Holistic Synthesis Systems-based Homeostatic	Worcester (1977), Shrivastava (1995), Gladwin et al. (1995), Hart and Milstein (1999), Ekins (2000), Lovelock (2000), Capra (2004), Zohar and Marshall (2004), Belz and Peattie (2009), Borland (2009a) and Guest (2010).

Management approach	Eco-efficient Socio-efficient Environmental management, sustainable development, ecocentric responsibility paradigm, 4Rs	Ecoeffective Socio-effective Ecological sustainability Waste equals food	Shrivastava (1995), Purser et al. (1995), Gladwin et al. (1995), Hart (1997, 2007), Hart and Milstein (1999, 2003), Starik and Marcus (2000), Banerjee (2002), McDonough and Braungart (2002), Stead and Stead (2004), Young and Tilley (2006), Borland (2009a) and Varey (2010, 2011).
Marketing strategy	Transitional Incremental Greenwashing Business as usual	Transformational Step change Ecologically sustainable	Gladwin et al. (1995), Menon and Menon (1997), Kilbourne (1998), Peattie (1999), Iyer (1999), McDonough and Braungart (2002), Banerjee (2002), Hart and Milstein (2003), Stead and Stead (2004), Young and Tilley (2006), Gronroos (2007), Kilbourne (2008), Belz and Peattie (2009), London (2009), Varadarajan (2010), Unruh and Ettenson (2010), Sharma et al. (2010), Varey (2010), Closs et al. (2011), Cronin et al. (2011) and Hult (2011).
Overall purpose	Human-centric, business as usual Sustainable development	Ecological sustainability Responsibility for all species and resources	Purser et al. (1995), Shrivastava (1995), Gladwin et al. (1995), Kilbourne (1998), Sharma and Vredenburg (1998), McDonough and Braungart (2002), Stead and Stead (2004), Porritt (2007), Borland (2009a) and Varey (2011).
Prospects for the future	Dystopian Destruction is the end game Only choice remaining is the rate of global destruction	Regenerative Restorative Systems-based Productive for business and nature Sustainable global society for all species, with the recognition of the need to reduce human population and consumption	Purser et al. (1995), Gladwin et al. (1995), Kilbourne (1998), McDonough and Braungart (2002), Du Nann Winter and Koger (2004), Stead and Stead (2004), Capra (2004), Zohar and Marshall (2004), Porritt (2007) and Borland (2009a, 2009b).

The anthropocentric epistemology also leads society to embrace a particular set of cultural values, metaphysical beliefs, institutions, habits and so forth, which collectively provide social lenses for interpreting the social world according to a dominant social paradigm (Kilbourne, 1998). There is no consensus on what constitutes the dominant social paradigm of Western industrial societies, but to dominate, it must be held only by dominant groups in society, not necessarily by a majority of people (Cotgrove, 1982).

Kilbourne (1998) cites two dominant social paradigm domains: The socio-economic domain, which incorporates political, economic and technological dimensions, and the cosmological domain, which refers to larger questions of existence, such as the structure (atomistic–holism), relation (domination–submission) and organisation (anthropocentric–ecocentric) of nature or the significance of nature itself. These background assumptions, largely unquestioned and/or unexamined, produce particular values, beliefs and behaviours (Kilbourne, 1998). We attempt to address some of these larger cosmological questions as a means to suggest an alternative direction for strategic marketing theory development.

3.2.2 *Ecocentric ideology*

Ecocentrism, broadly, is characterised by the belief that ecosystems have inherent worth for maintaining planetary homeostasis and all life. Through notions of holism, integration and synthesis, it asserts that human cultural systems must function within the safe operating limits dictated by ecosystems. Ecosystem integrity is paramount; animals and plants have as much right to exist as humans. It also establishes an underlying belief in the need for responsibility and stewardship towards plants, animals, wilderness and the planet (Dunlap *et al.*, 2000; Purser *et al.*, 1995).

The ecocentric epistemology is an alternative way of experiencing and evaluating the world, and it has acquired multiple names, including the new ecological paradigm (Dunlap *et al.*, 2000) and the ecocentric responsibility paradigm (Purser *et al.*, 1995), but it consistently represents a radical departure from anthropocentric epistemology. Ecocentric philosophers view anthropocentric assumptions as the root cause of environmental problems, so they express their explicit concern with emancipating ecosystems from the effects of human mismanagement, overuse and exploitation. To foster deeper appreciation for the intrinsic value of nature, ecocentrists seek to effect change at the levels of human beliefs, values, ethics, attitudes, behaviours and lifestyles. The relevant values align with movements to reduce human population growth and human consumption, preserve wilderness areas, protect the integrity of biotic communities and restore ecosystems to a healthy state of equilibrium.

Ecosystems are biologically diverse and ecologically sustainable; member organisms flourish in their respective niches, free from distress. This scenario allows for self-renewal, self-management and self-regulation in a dynamic, indefinite, self-perpetuating, closed-loop cycle (Borland, 2009a). Healthy ecosystems do not require repair, upkeep or management by humans, whereas unhealthy ecosystems demand environmental management, constant doctoring and engineering. King (1995) discusses the importance of avoiding ecological "surprises", activities initiated by humans that can destabilise ecosystems, and Rolston (1994, p. 71) realises that from an ecocentric ideology, the main issue is conserving natural values that do not put the health of ecosystems at risk, such that healthy ecosystems "produce natural values, as well as support human cultural values, and such productivity and support is the bottom line". This ideological shift places primary emphasis on the value of ecosystem integrity. Human cultural development can be encouraged if it sustains ecological integrity or ecosystem health (Diesendorf and Hamilton, 1997; Linnenluecke and Griffiths, 2010). That is, the focus is on ecological sustainability, rather than sustainable development or environmental management; ecological sustainability ultimately supports human existence (Bansal and Roth, 2000; Borland, 2009a; Porritt, 2007).

In Leopold's (1970) vision, humans evolve as they shift from an anthropocentric to an ecocentric ethic. Zohar and Marshall (2004) also claim higher levels of spiritual intelligence result from ecocentric and sustainability values, suggesting a much clearer meaning and purpose for human existence. The holism of the ecocentric epistemology emphasises the importance of the whole ecosystem, not individual members or parts, and removes humans as the sole locus of value. Such a radical change in beliefs, values and ethics can be psychologically challenging (Ketola, 2008; Naess, 1995), though the more rooted people become in understanding its principles, the more logical ecocentrism seems — to the point that anthropocentrism ceases to make sense. An ecocentric epistemology is not misanthropic (Gladwin *et al.*, 1995; Iyer, 1999) but rather amounts to an ideological and psychological, personal and collective shift, with a concomitant recognition of the physical constraints on individuals and organisational systems.

In Section 3.3, we consider how business and marketing strategy is guided by these two different epistemological ideologies which leads to the introduction of transitional and transformational strategies for ecological sustainability.

3.3 Ecocentric strategy development

Conventional definitions of strategic marketing and marketing strategy reflect an anthropocentric epistemology. Varadarajan (2010) distinguishes definitions of marketing strategy as either broad, with consideration of strategic resources and assets and their links to business and corporate strategy, or narrow, such that they focus on differences between marketing strategy and tactics. Yet, his definition of strategic marketing as a domain and marketing strategy as an organisational activity omits the essential role of the environment in providing natural resources and assets that are the source of all products and services. He acknowledges just that the "high level of interest among marketing academics and practitioners in sustainability-related issues is destined to have a significant impact on the nature and scope of the marketing discipline" (Varadarajan, 2010, p. 122). When examining the relationships among corporate strategy, business strategy and marketing strategy, he also suggests "strategic marketing decisions can be viewed as an organisation's decisions in the realm of marketing that are of major consequence from the standpoint of its long-term performance". This close relationship of the three strategy levels is essential for corporate success (Borland, 2009a), including an ecocentric corporate strategy.

Yet, strategy development that embraces ecological sustainability is virtually missing from corporate, business or marketing strategy literature, especially any approaches framed in ecocentric epistemology (Borland, 2009a; Dunlap *et al.*, 2000; Purser *et al.*, 1995; Shrivastava, 1995; Stead and Stead, 2004). Purser *et al.* (1995) place ecocentric theory development in its infancy stage and note that it is often regarded as unrealistic, though that perception may reflect the general lack of understanding of how to couple the science of ecological sustainability with the needs of commercial industry and human materialism. The challenge thus becomes to develop theory and practice that integrates the dualism of nature and human needs. Purser *et al.* (1995) claim that ecocentric theory development should proceed separately from existing anthropocentric theory development until it achieves sufficient legitimacy, coherence and maturity. They assert that the most urgent task at hand is assuring that the ecocentric responsibility paradigm enters into any formulation of organisational theory development and management practice and that organisation–environment relationships foster ecological sustainability.

Shrivastava (1995) argues that corporations have a responsibility to incorporate ecological sustainability into their logics, as an integral aspect of their effectiveness. Because corporations have the knowledge, resources and power to bring about enormous changes in the Earth's ecosystems, government policy and consumer behaviour in tandem could lead to true ecological sustainability. He also identifies benefits of ecological sustainability to corporations (Shrivastava, 1995), such as reduced long-term risk associated with resource depletion, fluctuating energy costs or product liabilities, as well as pollution and waste management. Yet Shrivastava (1995) recognises that a move to ecological sustainability requires an overall value reorientation in both society and corporations, from the current economic rationality to a broader ecological rationality focused on the long-term survival of all species (see Table 3.1, row 3).

In line with such theory, we focus on incumbent corporations and their role in ecological sustainability. Corporations are the primary engines of economic development (Gladwin *et al.*, 1995), with the financial resources, technological knowledge and institutional capacity needed to implement new strategies (Banerjee, 2002; Kilbourne, 2008). Examining ecological sustainability at organisational and functional levels is also necessary but underdeveloped, especially considering the scale of issues involved (Borland, 2009a; Kilbourne, 1998; Purser *et al.*, 1995; Shrivastava,

1995; Stead and Stead, 2004). We acknowledge though that corporations are only one ecological sustainability gap; consumers and governments must be willing to participate too, but a discussion of these two groups is beyond the scope of this chapter.

3.4 Ecocentric business strategies

We propose two strategic alternative approaches to ecological sustainability. Our transitional strategies (Table 3.1, column 2) maintain an anthropocentric epistemology and the dominant social paradigm, as can be easily identified in the modern corporate arena. They are characterised by 5Rs — *reduce*, *reuse*, *repair*, *recycle* and *regulate* (see Table 3.2 for details). They are also associated with the adoption of eco-efficiency and socio-efficiency management (McDonough and Braungart, 2002; Young and Tilley, 2006). These transitional strategies are linear, cradle-to-grave, open-loop, dualistic approaches that create continuous improvement and incremental change (Hart and Milstein, 1999), driven by a desire for competitiveness or differentiation. They do not, however, encourage natural diversity, creativity or productivity. In this sense, they represent anti-sustainability and act merely to slow down the eventual death and destruction of resources and habitats of which corporations, consumers (citizens in the ecocentric view) and government are stakeholders.

Transformational strategies (Table 3.1, column 3) instead embrace ecocentric epistemology and the ecocentric responsibility paradigm (Purser *et al.*, 1995). They are characterised by an alternative 5Rs — *rethink*, *reinvent*, *redesign*, *redirect* and *recover* (see Table 3.2 for details) (Borland *et al.*, 2016). By working within the constraints of natural ecosystems, transformational strategies incorporate ecoeffectiveness and socio-effectiveness (McDonough and Braungart, 2002; Young and Tilley, 2006) and represent holistic, cradle-to-cradle, systems-based, closed-loop, visionary approaches that create discontinuous change and creative destruction (Hart and Milstein, 1999). They can be competitive but achieve better firm performance through collaborative, innovation-oriented strategic alliances (Child *et al.*, 2005; Hall and Vredenburg, 2003; Nidumolu *et al.*, 2009; Porter, 2008; Porter and van der Linde, 1995a, 1995b; Pujari *et al.*, 2003, 2004; Seitz and Peattie, 2004; Senge and Carstedt, 2001; Sharma *et al.*, 1999; Slater *et al.*, 2007).

Table 3.2. Transitional and Transformational 5Rs

Transitional 5Rs	Transformational 5Rs
Reduce Reduce the quantity of material used in manufacturing certain products, and domestically reduce the quantity/number of products used.	**Rethink** This first stage requires completely rethinking the concept of what the product is: Is a car a car, or is it a means of getting from A to B? After determining the function of the product, we can think of different ways to satisfy the function in an environmentally, closed-loop way.
Reuse Wherever possible, reuse materials and products so that the overall volume demanded is reduced and the product is used to its fullest extent.	**Reinvent** Make way for reinvention: This creative, innovative, brainstorming process identifies completely new concepts that may or may not be based on existing products. Alliances and clean technology may be required.
Repair Some products can be repaired and reused, rather than being disposed of, thus extending their useful life and reducing demand for new products.	**Redesign** Once new concepts have been identified, redesign needs to embrace ecological requirements as its primary position so that products (and services) are designed to be made from biological material or technical materials only, eliminating waste and toxic residues. For example, an upcycled vehicle might run on water and release no dangerous residues, but instead contribute positively to the environment by cleaning air or water as it runs.
Recycle If a product cannot be repaired or reused, recycling options exist, whether domestically, municipally or through a corporate recollection scheme. Conventional recycling is a finite process for most products, because the elements ultimately become degraded beyond usefulness, leading to downcycling and disposal in a landfill or incineration.	**Redirect** Redirect and recover affect the product at the end of its life. Redirect refers to the need to have two clear channels for waste materials: One where *all* waste materials go back into the industrial cycle so that nothing is wasted and pollutants are not released to damage the environment, thus creating a closed loop, and another for biodegradable materials that can go back to nature without causing any physical or chemical damage. These two channels need to be kept separate. Sophisticated, productive, profitable channels need to exist to make it a reality, so that industrial materials can be infinitely cycled without loss of quality. This step also addresses the increasing scarcity of some raw materials (e.g. copper).
Regulate Increasingly companies and individuals are subjected to laws, restrictions and regulation that control activities associated with waste material. These restrictions are set to increase in the future.	**Recover** To recover scarce (and not so scarce) elements and materials and use them in new production and market opportunities, thus maintaining their market value (industrial symbiosis) and again closing the loop. This cycle then operates as an infinite, circular system with no end. Only increases in end-user demand generate the need for virgin resource extraction.

Another quality sets transformational strategies apart from other strategic approaches: They work with nature rather than against it and thus require significant scientific, psychological, and strategic understanding by the focal company. The progression towards a transformational strategy is not necessarily smooth and may require a step-based change in identity and leap of faith. Just as transformation at an individual level requires a fundamental shift in the depth and level of the individual's learning and understanding, usually precipitated by a negative, life-changing experience (Borland, 2009b; Du Nann Winter and Koger, 2004; Zohar and Marshall, 2000, 2004), at the collective, corporate level, the experience is often equally life-changing for the very orientation of the company. The PVC manufacturer Hydro Polymers, for example (Leadbitter, 2002), experienced the threat of closure from the significant negative publicity it suffered as a result of some of its activities before it changed to a transformational business strategy for sustainability.

Understanding transformational business strategies is central to developing strategies for ecological sustainability, but our recognition of two macro-business strategies suggests some key questions are as follows:

1. Are transitional and transformational business strategies mutually exclusive or progressive?
2. Is a transitional strategy likely to become embedded in a firm, such that it can no longer progress to a transformational strategy?
3. Does a transformational strategy require the company to go through a transformation?

Transitional and transformational strategies are not mutually exclusive; a company might initially adopt a transitional strategy that encourages eco-efficient behaviours, such as the 5Rs, to introduce employees, suppliers, customers and other stakeholders to new attitudes, values and behaviours. In the process, the leap to a transformational strategy becomes easier; if the firm applies transitional behaviour to ecoeffectiveness, for example, using only biological and technical nutrients (which we define subsequently) (McDonough and Braungart, 2002), the difference likely becomes transformational. A transitional strategy thus can be a first step towards a transformational strategy.

Yet, it is also possible to adopt a transformational strategy without even introducing a transitional strategy; equally, some firms may become stuck in a transitional mode without even progressing to a transformational strategy. An embedded transitional mode is dangerous (Hart, 1995), comparable to simply continuing with business as usual in a conventional business and marketing mode.

Finally, the process of transformation is usually led by an individual within an organisation, rather than the organisation *per se*. Ecocentric transformational leadership is a central element of the success of corporate ecological sustainability (Borland, 2009a; Closs *et al.*, 2011; Stead and Stead, 2004; Zohar and Marshall, 2004). Its further discussion is beyond the scope of this chapter though.

Section 3.5 begins with the process of defining what ecocentric business and marketing strategy solutions will look like from a theoretical standpoint.

3.5 Ecocentric business and marketing solutions

Ecocentric business and marketing solutions evolve from ecocentric business strategies. In particular, transformational business strategies operate most effectively at an industry level and then at the individual company level (Bansal and Roth, 2000; Hart and Milstein, 1999). By creating an ethos of cooperation, collaboration and innovation across competing firms, it can support future industry development and survival (Child *et al.*, 2005; Hart, 1995; Nidumolu *et al.*, 2009; Shrivastava, 1995). McDonough and Braungart (2002), Stead and Stead (2004), Hart (1997, 2007) and Hart and Milstein (2003) have contributed to the development of ecocentric business and marketing solutions, though none of them frame their work according to an ecocentric epistemology. Their approaches are varied, but in combination, they reflect a logic and flow that makes strategic sense.

McDonough and Braungart (2002) assert that the main issue for marketing ecologically sustainable products and services is the design of physical products. Many household products give off high levels of noxious and dangerous substances, and industry disposes of vast amounts of dangerous chemicals that are crippling the environment and harming human health (McDonough and Braungart, 2002). These authors conclude that industrial ecosystems should enhance and add to their local environment, rather than poisoning the environment and human health. Their mantra illustrates that product conception, development, manufacture, use and disposal must follow holistic, systems-based, closed-loop principles that prevent pollution and waste.

Using the term ecoeffectiveness, McDonough and Braungart (2002) also suggest that all products should be produced from two types of materials: Biological nutrients and technical nutrients. Biological nutrients biodegrade and can be returned to the biological cycle without inflicting any damage; technical nutrients do not biodegrade but can be circulated continuously through the industrial cycle, which eliminates waste and pollution and reduces resource use. This more positive outlook for the future could allow humans to continue with their current lifestyles and quality of life. In providing this cradle-to-cradle, closed-loop, transformational approach to product designs, strategic marketing and corporate vision, McDonough and Braungart (2002) also offer a transformational solution to managing supply chains.

In another example, Stead and Stead (2004) emphasise three value chain models. Type I is a conventional, cradle-to-grave supply chain; type II is a transitional version that incorporates recycling activity; and type III, of most interest to ecocentric business and marketing, depicts a cradle-to-cradle, closed-loop, value chain in an open living system economy, with no waste or pollution. Using only biological nutrients, renewable energy sources and technical nutrients in industrial systems, it produces safe biological waste that get reabsorbed into the biological system. However, they recognise that a supply chain using only biological or technical nutrients needs new manufacturing technology and processes, including clean technologies that are yet to be developed (Hart, 1997; Hart and Milstein, 2003).

Government support for such transformations will be essential, coupled with industrial collaboration. However, the current global financial predicaments, turbulent nature of major industrial economies and exponential population growth in developing and emerging economies suggest that it may also be a way out of economic and environmental devastation, in a more positive and life-enhancing way than has been proposed previously. The transformation to this type of system is challenging and requires major research and development investments.

The contributions of McDonough and Braungart (2002) and Stead and Stead (2004) can be combined with the recommendations of Hart (1997, 2007) and Hart and Milstein (2003). Their sustainability portfolio matrix offers companies a roadmap or vision for sustainable change in four stages: Pollution prevention, product stewardship, clean technologies and sustainable vision. The stages are progressive and offer increasing challenges to a firm's ecocentric strategic marketing activity as it moves towards a sustainability vision. Pollution prevention and clean technologies affect the internal operation of the company; product stewardship and sustainable vision also engage external elements, such as suppliers, customers and other stakeholders. Even in the first stage, the process can be difficult to implement at a practical level, because it demands not just pollution reduction (which would be transitional) but pollution elimination. Therefore, to enter the portfolio at all is a challenging task, aligned with McDonough and Braungart's (2002) ecoeffectiveness or Stead and Stead's (2004) type III value chain.

In integrating these three closed-loop, systems-based, transformational, ecocentric epistemological approaches, we offer a vision of ecocentric business and marketing solutions that incorporates the redesign of products (and services) using only biological and technical nutrients as components and materials. Such usages enable the supply chain — from source to consumer and beyond — to close its loop and avoid leaking or leaching any unwanted or dangerous substances into the environment. It also closes the waste loop, such that the transfer of energy and nutrients represents a continuous process from cradle to cradle (type III). In Hart's (1997) sustainability portfolio matrix, type III succeeds in achieving pollution prevention and product stewardship; it also requires clean technologies to be developed and provides a clear and unambiguous ecological sustainability vision that could be applied to firms in developed nations, emerging economies, developing nations and base-of-the-pyramid societies worldwide (Hart, 2007; London, 2009; Prahalad and Hart, 2002).

In Section 3.6, we begin to map out what an ecocentric marketing strategy vision might look like and provide case examples and a managerial tool to help academics and practitioners identify ecocentric marketing strategy. Section 3.7 of this book embraces the general applicability of ecocentric business and marketing strategies by offering six universal foundational premises, as part of future conceptual and theoretical development. Sections 3.6 and 3.7 together represent the primary contribution of this book.

3.6 An ecocentric strategic marketing vision

The preceding sections provide key input to inform ecocentric marketing strategies, which should lead to an ecologically sustainable approach to product conception and design that ensures the outputs do not damage the environment or people. Such strategies also should close the loop in supply chains, changing or eliminating the notion of waste; offer opportunities for product differentiation and thus competitive advantage; and provide a vision of what a truly ecologically sustainable society would look like, filled with ecocentric products and services.

Ecocentric marketing strategies are transformational in nature and follow the format of transformational business strategies (Table 3.1, row 7): They are ecologically sustainable and pursue an ecoeffective, socio-effective route that is closed-loop, cradle-to-cradle and systems-based. They encourage health and abundance for all species, because no damage gets inflicted on the physical environment or health.

If a company adopts an ecocentric transformational marketing strategy, its sphere of influence should extend to consumers, suppliers (Sharma and Vredenburg, 1998) and other firms in the same industry, creating leadership and first-mover advantages (Unruh and Ettenson, 2010). At a strategic level, firms must first realise the differences between transitional and transformational approaches, then make a conscious decision to adopt a transformational strategy. Transitional strategies may provide a useful first step, but they are also tantamount to greenwashing and maintain a destructive, business-as-usual approach. It is therefore essential that a company embraces a vision to move beyond transitional strategies if it genuinely aims to contribute to the survival of all species and the environment and hopes to make its strategic marketing activity part of the solution rather than part of the problem. We therefore introduce some practical steps for implementing ecocentric transformational marketing strategies.

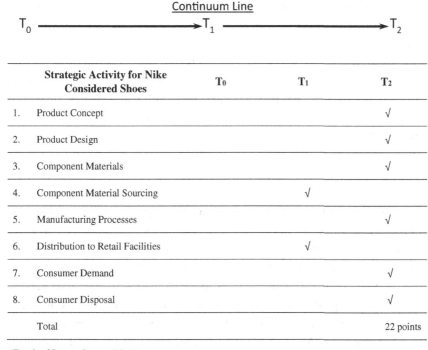

Strategic Activity for Nike Considered Shoes		T_0	T_1	T_2
1.	Product Concept			√
2.	Product Design			√
3.	Component Materials			√
4.	Component Material Sourcing		√	
5.	Manufacturing Processes			√
6.	Distribution to Retail Facilities		√	
7.	Consumer Demand			√
8.	Consumer Disposal			√
	Total			22 points

Total = 22 (out of a possible 24).
(A transformational activity = 3 points, transitional activity = 2 points, traditional activity = 1 point.)

Figure 3.1. Ecocentric Transformational Business and Marketing Strategy Grid

It is important to remember that the move from conventional through transitional to a transformational strategy can be stepwise. If we refer to a conventional strategy as T_0, a transitional strategy as T_1 and a transformational strategy as T_2, we might depict a continuum, with T_0 at one end, T_2 at the other and T_1 in the middle of the line (Figure 3.1). Or we could create a grid, with T_0, T_1 and T_2 across the top and the different strategic marketing/business activities along the side (Figure 3.1). For each product, we can then assign individual activities, such as product designs, to conventional, transitional or transformational categories. This grid produces a numerical score that reflects the status of each product. With this progressive, aspirational approach to strategic marketing, marketing departments and companies gain the opportunity to assess their progress towards transformational strategies and evidence to bolster their claims that they are working towards ecological sustainability in a genuine and life-enhancing way. Of course, firms are unlikely to adopt any strategic approach that does not provide secure financial returns. Ecocentric transformational strategies change the very nature of the product being sold, such that they can enhance financial returns through genuine sustainability-based differentiation and competitive advantage.

3.6.1 *Case examples*

A review of existing companies does not turn up any company that has completely adopted an ecocentric transformational strategic approach, though many companies are making genuine attempts to be sustainability-led. A particularly interesting example, considering its prior transgressions, is Nike. Its line of casual shoes, Nike Considered, appear to follow an ecocentric approach in that they are made of vegetable-tanned leather, which eliminates toxic chromium (traditionally used to tan leather) from the waste pipeline. After its product usage, the leather will decompose naturally in compost heaps to become food for other species (biological nutrient) and leave no toxic residues. The soles of the shoes are made from recycled rubber and are infinitely recyclable if returned to the company (technical nutrient). Because there are no adhesives involved in constructing the shoes, production workers in factories and the environment experience no toxic effects. The components of the shoe are designed to "pop" together and can be completely disassembled for easy recycling or reuse. The shoes are desirable to consumers, and the demand is high. Finally, all their materials are sourced within 200 miles of the factories that produce them to reduce fuel consumption. These shoes thus score high in Figure 3.1 on the product concept, product designs, component materials, manufacture and consumer demand categories; their score is somewhat lower for component and retail distribution. Although it represents only one product line at this stage, Nike considered shoes provide interesting evidence that an international conglomerate can experiment successfully with ecocentric models and strategies.

Herman Miller, the office furniture manufacturer, has developed a range of office chairs that follow similar principles. The seats are made of fabric constructed solely from biodegradable materials; if added to an aerobic composting environment, they will biodegrade naturally and leave no toxic residues. The frames and plastic components of the chairs are constructed such that they can be disassembled, recycled or reused indefinitely, without down-cycling, in a closed-loop industrial cycle. Thus, they create no waste and eliminate the need for virgin raw materials. Closing the manufacturing loop changes the emphasis on the value of component parts. At the end of their life, instead of products being regarded as waste to be sent to landfill, manufacturers become highly interested in their return, because they are the input for the next round of production. The components are valuable raw materials for new products, which make the relationship between the manufacturer and material more positive. This strategy can also induce changes in the marketing strategy, such that the firm might become more interested in renting products to consumers rather than selling them, to ensure it receives the products back at the end of their productive life in a particular iteration (McDonough and Braungart, 2002).

Transitional and transformational strategies also represent an opportunity to adopt existing sustainability-oriented management tools that are more widely used. In particular, lifecycle analysis and biomimicry are examples of management tools that can be applied at either a transitional or transformational level; the difference being whether each is used in a cradle-to-grave or a cradle-to-cradle fashion and thus whether each is conducted as a closed-loop exercise or not. Closed loop, cradle-to-cradle lifecycle analysis and biomimicry are tools that can be used within transformational strategies. Open-loop, cradle-to-grave lifecycle analysis or biomimicry are tools that would fit with transitional strategies. Therefore, lifecycle analysis and biomimicry have the potential to fit either strategy type.

3.6.2 *Ecocentric strategic marketing premises*

Varadarajan (2010) identifies some foundational premises for marketing strategy, such that to be universal, they must be generalisable across products, markets and time horizons. He also articulates two key purposes of a marketing strategy: To enable a business to achieve and sustain a competitive advantage and to influence consumers' preferences. We add another purpose: A marketing strategy must incorporate the physical environment as the source of physical well-being for all species, as well as the source of all products and services. Marketing strategies then become grounded in physical, scientific reality as well as human social reality. They will thus become more stable and sustainable, in both commercial and ecological senses.

Ecocentric marketing strategies meet Varadarajan's (2010) foundational premises to provide a competitive advantage, create organisational assets, nurture exchange relationships, influence consumers' purchasing behaviour, leverage new points of differentiation and enhance the salience of non-price criteria. We propose six additional universal premises, grounded in environmental and ecological science, to which ecocentric transformational marketing strategies must also adhere are as follows:

1. Adopt the design, manufacture, consumption and disposal of ecoeffective products and services.
2. Utilise energy from renewable resources such as solar and bio-gas, at both commercial and domestic levels.
3. Engage in habitat reconstruction and the preservation of and respect for all species.
4. Educate people about their individual responsibility towards the environment and other species.
5. Seek financial investments from governments that support ecoeffective industry, firm and product development for future economic stability, collaboration and competitiveness.
6. Promote sensible family size worldwide, with no more than two children per family, and support the adoption of orphans.

In addressing these premises, we propose a definition of ecocentric transformational marketing strategies:

> Companies that satisfy the needs of industrial and consumer markets within biophysical constraints, only exploiting resources at a rate at which they can be sustainably maintained, recovered or replenished in cradle-to-cradle, closed-loop ecological systems.

From this definition, it is also possible to summarise key ecocentric transformational marketing strategy principles as follows:

- Product design and innovation from ecological core competencies.
- Value from sustainability values.
- Competitiveness from ecological stability.
- Collaboration from shared sustainability goals.
- Solutions from shared sustainability understanding.
- Vision from ecocentric marketing leadership.

In Section 3.8, we reiterate our original research question and examine next steps for research and management. We conclude by reflecting on the approach taken and the objective of finding new research linkages between strategic marketing activity and ecological sustainability.

3.7 Managerial relevance and further research

It is possible to reconsider our original research question: Can strategic marketing be reconceptualised to reflect ecocentrism and ecological sustainability? We answer our question in the affirmative, which suggests the need to consider its relevance and opportunities for managerial practice and research. At a broad level, we offer managers and researchers a clear, easy-to-apply approach for categorising sustainability strategies. Much ambiguity persists as to what constitutes a strategy for sustainability and what does not. We propose a foundational conceptual framework that will allow researchers and managers to identify quickly whether a business is adopting a transitional or transformational sustainability strategy and thus how the firm might be guided to develop its strategy further. Auditing their current position in this way also enables firms to move forward. Identifying the path from transitional to transformational sustainability strategies will help firms and their different functions produce a sustainability vision, mission, values, goals and objectives.

The clear delineation of eco-efficiency and ecoeffectiveness in transitional and transformational strategies also may support the future development of a universal "traceability mark" for products and services that claim to be sustainable. Firms could differentiate themselves according to their actual ecological sustainability credentials. As consumers become increasingly discerning and information savvy regarding products' provenance, such a logo, through becoming a trust mark, could provide a source of competitive advantage and improve corporate reputation.

Finally, firms and their functions can identify themselves clearly according to the two sustainability strategies. They then may seek out likeminded partners as suppliers, distributors, retailers and so forth, in their network of operations. In particular, firms following a transformational sustainability strategy can create transformational networks of firms.

At a more focused level, this research provides opportunities to examine a firm's approach to the sourcing, product designs, manufacture, distribution, usage and disposal of particular products. Each supply chain can gather evidence of the existence of a transitional or transformational strategic approach to the creation of products and the elimination of carcinogens, mutagens and persistent and accumulative environmental toxins. In turn, opportunities arise to examine, in detail, whether each firm tends to adopt and then persist with a transitional or transformational strategic approach or if it is possible to move from one to the other, and which mechanism enables such shifts.

A wealth of research opportunities thus emerge for academics to establish, both qualitatively and quantitatively, whether firms adopt either sustainability strategy, what their characteristics are, how successful firms have been and what their future plans are. Further theoretical research also should test the application of the ecocentric universal premises and principles and both sustainability strategy types. Thus, the managerial relevance and research opportunities associated with ecocentric transformational marketing strategies are significant. In our view, this new approach to strategic marketing opens a new area of inquiry and suggests productive avenues for research and management practice in the coming years. Finally, further research could look towards recent manifestations of an anthropocentric ideology, terminology such as anthropocene and resilience that suggest a different view on ecosystems and human–nature interactions. Whilst the focus here has been on the epistemological debate and thus having left out discussion of geological time zones, we believe that the suggested research avenue potentially could add to the grounding for the epistemological endeavour of this chapter.

3.8 Conclusions

This chapter has achieved several key outcomes. First, we advance the approach presented by Purser *et al.* (1995) by identifying business and marketing strategies that reflect ecocentric epistemology and ecological sustainability. Although still descriptive at this stage, it represents, to the best of our knowledge, the first attempt at such an identification.

Second, we have adopted Kilbourne's (1998) cosmological domain to address some of the larger questions of existence — such as the significance of nature, as well as its structure (atomistic–holism), relation (domination–submission) and organisation (anthropocentric–ecocentric). We take the perspective of ecocentrism to examine marketing and sustainability and thus challenge some existing beliefs, values, ethics, attitudes and behaviours that are pervasive in corporate and consumer society. In so doing, we illustrate an alternative way to achieve ecologically sustainable directions in future strategic marketing activity.

Third, we follow the guidance provided by Varadarajan (2010) in defining the foundational premises of strategic marketing. Ecocentric transformational marketing strategies are consistent with Varadarajan's (2010) recommended list; this research also has identified six universal premises and six strategic principles that are uniquely applicable to ecocentric transformational marketing strategies.

Fourth, this paper reveals the linkages, and thus the bigger picture, between marketing strategy and sustainability and offers a means for studying ecological sustainability as an academic topic in a business school or commercial context. Framed within the ecocentric epistemology, our work connects business strategy with marketing strategy with sustainability, then defines the relationships among the three through the application of an ecocentric, transformational, cradle-to-cradle, systems-based, closed-loop approach. We thus illustrate a new foundational link between marketing and sustainability and define them as connected subjects for further strategic marketing research enquiry and management practice.

Acknowledgment

This chapter is based upon two previously published articles by the authors, namely Borland, H., Ambrosini, V., Lindgreen, A. and Vanhamme, J. (2016). Building theory at the intersection of ecological sustainability and strategic management. *Journal of Business Ethics*, Vol. 135, No. 2, pp. 293–307 and Borland, H. and Lindgreen, A. (2013). Sustainability, epistemology, ecocentric business and marketing strategy: Ideology, reality, and vision. *Journal of Business Ethics*, Vol. 117, No. 1, pp. 173–187.

Chapter 4

TOWARDS AN
ECOLOGICAL ACCOUNTING

Towards an Ecological Accounting:
Tensions and Possibilities in Social
and Environmental Accounting

Rob Gray

University of St Andrews, St Andrews, Scotland
rhg1@st-andrews.ac.uk

4.1 Introduction

Ecological accounting is potentially a very broad and diverse term. At its most precise it refers to accounts, narratives, reports about ecological systems themselves; their functioning, their health and the threats they face. The term can be used to encompass accounts about regions or about biota or even species. Ultimately, if understanding of ecology is our aim (assuming such a rationalist objective is tenable) then these are the accounts to which we need to pay the greatest attention (Russell *et al.*, 2017). As far as one can tell, it is not fanciful to imagine that human societies with a closer relationship with their environment and a less rabid attachment to development naturally enjoyed shared accounts of their forest, their water systems, the wildlife and so on as an essential element of their lived experience (Leopold, 1949/1970; Passmore, 1980; Tucker, 1999). In modernity, such accounts are the business of specialists in geographical and biological sciences and are, inevitably, somewhat remote from the daily lived experiences of most of us. Although accounts are generated, passions aroused, and resources mobilised within modern experience on specific issues — on such matters as the plight of the bee, concern for tigers or, for example, the hunting of whales — these attempts to introduce more closeness between the modern experience and the essence of ecological appreciation are only ever partially successful (Birkin and Polesie, 2012).

If one thinks more widely about the notion of an ecological account, then we find ourselves including the increasing variety of global snapshots of the planet. Global accounts have, arguably, been more influential and their effects long lasting than the more specific accounts. Such general accounts vary from the seminal texts such as Leopold's *A Sand County Almanac*, (Leopold, 1949/1970), Goldsmith's collaboration on *Blueprint for Survival* (Goldsmith *et al.*, 1972) and Schumacher's (1973) *Small is Beautiful* through to the range of substantive records of the planet's health including such things as WWF's *Living Planet Report* (see, for example, WWF, 2016), UNEP's *Global Environmental Outlook* (see, for example, United Nations Environment Programme, 2012), the Millennium Eco-Assessment (United Nations Millennium Ecosystem Assessment, 2005) as well as periodic bulletins such as the Worldwatch Institute's *State of the World* series (see, for example, Brown and Flavin, 1999). Speaking personally, it is these accounts which form an essential part of the backdrop and context to how I understand the relationships between modernity, international financial capitalism, human society and ecology. It is these accounts which raise concerns over, for example, the (non-) accountability of the multinational corporation for matters of societal injustice and conflict. It is these accounts which are in large part responsible for generating my sense of outrage and aesthetic disgust at the failures of modernity (Gray, 2010). It is such matters, albeit alongside accounts of societal injustice (World Inequality Labs, 2018) that provide the various manifestations of (the admittedly more narrowly focused) social and environmental accounting with its motivation, its potential and its importance.

It is crucial to note that social, environmental, ecological and sustainability accounting is in no sense a coherent area of theory or practice. Its subject matter is approached from so many different points of view and embraces a very wide range of intentions and purposes (Gray *et al.*, 2014). Consequently, the distinction I will draw here between ecological and social/environmental accounts is almost certainly somewhat arbitrary and rather artificial. Ecological accounts should, quite obviously, be accounts of ecology and ecological systems. The predominant (though not exclusive) focus in social and environmental accounting as generally understood (Gray *et al.*, 2014) is upon organizations and institutions rather than upon ecological (or social) systems as such. Indeed, as Russell *et al.* (2017) so explicitly remind us, there is very little society or ecology in most social and environmental accounting practice (and theory) itself (p. 1436). Consequently, if social and environmental accounting is what I can speak about (and I think it is), there is a danger that I have nothing to say about ecology. Equally, were I to talk exclusively about ecological accounts, I would find myself out of my depth and ill-equipped to speak about biological and geographical sciences. However, if (as I maintain) social and environmental accounting are (or should be) entirely motivated by ecological and societal information, anxiety and outrage then we can explore such accounts to the extent that they help us make any sense of the abuses and desecrations of our times. Crudely, if modernity and international financial capitalism are the primary sources of assault upon society and ecology, then the focus of social and environmental accounting is (or at least should be) upon those sources and how they might be undermined, mitigated and eventually removed.

As a result, it is an ecologically and societally informed exploration of social and environmental accounts — in their broadest sense — which will be informed in this chapter.

This chapter is organised as follows. Section 4.2 offers a brief reflection upon the wider question of the giving and receiving of accounts before I explore in Section 4.3 what the academic and practitioner experiences of accounting might be able to bring to our understanding of ecological accounts. Section 4.4 seeks to reflect upon how social and environmental accounting might help us navigate from modernity to a world of intrinsic sustainability. In Section 4.5, I provide a very brief review of just a few of the different attempts to approach and/or articulate an ecological accounting and then Section 4.6 provides a reflection on the elements of this essay and some tentative conclusions.

4.2   The ubiquity of "accounting"?

Accounts, it seems, are a ubiquitous part of human experience. As far as one can tell, regardless of time or place, human beings give and receive accounts as an inevitable part of who and what they are. Accounts of culture and of brave deeds, accounts of their origins and cautionary tales; stories, histories and the best way to acquire food, are so embedded in human intercourse that, I admit, I find it very difficult to imagine humans without accounts. This is not to say that accounts are necessarily objective, or neutral or benign, (a subject I return to later), but it is to suggest that blanket criticisms of accounting (in whatever form) are not very helpful. That is, the notion that in some perfect state of grace where mankind is, for example, at one with the Earth there would be no accounts at all is, I fear, entirely fanciful and oxymoronic. So, somewhat belabouring the point: if you do not have accounts in some form or other, you probably do not have human beings — and whilst such an extinction might have its attractions (Gray and Milne, 2018) — it does rather suggest that the more extreme critiques of accounting broadly as (say) unnatural or essentially malign are incorrect.

At the same time, it is diverting to try to imagine whether non-human species account to each other. It does seem to be stretching a point to suggest that bees' communication of a location of pollen or the alarm calls of geese or the communications between dolphins are actually accountings. It is not obvious that such a stretch is actually very helpful either. What does seem interesting is the inference that accounting is one of those things that appear to differentiate humankind from all other species. This may not be very earth-shattering an insight but at a minimum it seems there is value in differentiating the species without falling foul of the pejorative sense of "being anthropocentric".

This preamble is necessary, I think, because there have been serious challenges that claim one should not be accounting for ecological issues at all (cf. Cooper, 1992; Cooper *et al.*, 2005, etc., see also, Gray *et al.*, 2017). This does, I fear, misunderstand the different ways in which one might account and the different purposes for which it might be under-taken. In the first place, if humankind is essentially an accounting animal, then the species may well account for all kinds of things — including, of course, ecological matters as the very basis of its context and existence. However, there are accounts whose purpose might well be considered essentially malign as well as those accounts whose use is predominantly anti-ecological. This concern is at its most piquant when we consider economic accounts — the stock-in-trade of a conventional accounting. Such accounts are, by definition, unnatural (in that they explicitly exclude all but the economic) and, designed as they are to facilitate and encourage ecological (and social) exploitation, they are also malign in most circum-stances, (see, for example, Thielemann, 2000).

So I shall take it as given (at least for the purposes of this essay) that there is nothing *essentially* problematic with the giving and receiving of accounts — after all, we all need inspirational accounts of possibility — but in order to explore how we might imagine exploiting a benign ecologi-cal accounting we need to constantly and carefully distinguish the kind of accounts of which we are speaking *and* the purpose and intentions of those accounts. As I draw from the literature of accounting in my reflec-tions, it is with accounting that the next section is concerned.

4.3 Accounting and where to draw its parameters?

Conventional accounting — the arena of expertise and practice normally associated with accountants — has little or no natural capacity for environ-ment, ecology or, indeed, society. This instinctive, deeply rooted insistence on excluding all that might more naturally be thought of as good about humanity and its planet does not mean that accounting has no impact in those arenas. Quite the opposite: accounting practice is brutally implicated in both social and environmental degradation. As Johnson (2017) notes:

> … human beings have long acted as destroyers, not stewards, of Earth and its inhabitants. I believe that an extremely important contribution to this unfortunate treatment of Earth are the financial practices of account-ants, economists, and business leaders (p. 168)

Birkin (1996) picks up this theme and goes further to enquire

> why didn't accountancy embrace ecological holism in the first place?
> A holistic understanding has not informed the accounting pathway
> followed to date (p. 233).

But accounting never did recognise, let alone embrace, holism and it continues to practice as predominantly a manifestation of modernity and of the thinking that frames the world through neo-classical economics. It is largely this joint focus on modernity and economics which ensures that as along as humanity insists on current forms of accounting then, consequently, environmental and societal devastation are largely inevitable (Gray, 1990, 1992; Gray *et al.,* 1993). The point of this is that until such time as current practices are removed, reformed or made irrelevant, the prospects for "developed" and "developing" societies to be able to join an ecologically centred mankind that knows to nurture and to live within ecological (and societal) boundaries seem exceedingly unlikely. Until then, Andre Gorz's observation about accounting knowing all about more and less but nothing of enough remains a crucial indictment (Gorz, 1989).

Faced with this malign accounting, there has grown up a broad area of enquiry (and, to a degree, practice) which has sought to address this state of affairs. Very broadly, social and environmental accounting has been concerned to:

(a) Identify and locate the malign factors within current accounting practice and theory (and, consequently, identify and locate the implications for finance and for business).
(b) Seek out ways in which these practices might be mitigated or even eradicated.
(c) Develop new ways of accounting for different things for different purposes.

The point being that current conventional accounting is only a very small subset of all the possible ways in which humanity can — and does — give and receive accounts. Equally, alternative forms of account can expose both the partiality and misleading nature of the dominant financial accounts. This places social and environmental accounting right in the heart of the power struggle over the dominant narratives of modernity. That is, accounting tends to be concerned with organisations and institutions — and typically with the power of such entities. Conventional accounting typically seeks to maintain the grip of those with the power in modernity. Social and environmental accounting on the other hand typically: seeks to exploit ways in which the economic dominance can actually be understood as benign; looks to demonstrate how accounting can be reformed to achieve more benign outcomes for society and the planet; or it sees its role as one of the challenges and dissonances — as "speaking truth to power" (Gray *et al.*, 2014).

There is little doubt that dominant accounting and indeed a great deal of social and environmental accounting practice are concerned with maintaining a status quo which demonstrably has nothing to say about a just society or a nurtured ecology, but there is a major strand of social and environmental accounting scholarship that draws from wider, more global accounts in order to seek to establish mechanisms by which the dominant organisations and institutions could be explicitly and manifestly held accountable for their interactions with society and the planet. It is this sense in which I want to talk about social and environmental accounting as a precursor to a world of ecological accounts.

It is quite apparent that our world of modernity and international financial capitalism is destroying nature as both the component of humanity that gives us meaning and, more instrumentally, the very basis upon which our continuation depends. We are clearly a long way from any utopian position — a "state of grace" as Levitas (2013) calls it (Gray and Milne, 2018). Dispute, very properly, then coalesces around questions of whether — and if so how — we might get from one to the other. Accounts are inevitably one part of any possible movement as they offer: expositions of the unacceptable consequences of the present; explanations of root causes; and the imagining of future possibilities and of utopian visions. Accounts alone will clearly not be enough, but they can serve to arouse, to warn, to shock, to guide and to enlist. Such accounts, as we have seen, can range from the very personal, the rhetorical, the poetic and the emotional to the objective, rational, descriptive and analytical: and every point in between.

It is a central tenet of much social and environmental accounting (see, for example, Owen, 2008) that the organs and institutions of modernity must be directly and bravely confronted and challenged: that the organs of the institutionalised violence against the planet and society (however unintended and seemingly remote that might be) are firmly held accountable for their impacts on society and the environment. Only, it might be argued, by exposing the essential contradiction in western comfort and privilege might there be any chance of (drastic) reform being even considered.[3] In essence, this approach to social and environmental accounting embraces the moral imperative to puncture what Adams refers to as "hypernormalisation" — where the outrageous becomes increasingly accepted as normal and unexceptional; where instead of dealing with climate change, species extinction and so on we are instead "admitted to the fantasy land of accelerated mobility and consumption" (Adams, 2016). This hypernormalisation is one explanation as to why we appear to be unable to begin to address new and better ways of organising. The formal accountability of organisations and institutions — especially the accountability of the largest organisations and institutions — can therefore be seen as an essential precursor to the transformation that Birkin (1996), for example, suggests, any sustainable development will require. Namely that:

> development decisions are motivated and informed primarily by the requirements of adapting to the needs of the ecosystem and, hence, that industrial practice develops in accordance with an integrated ecological and economic understanding of efficiency. (p. 232)

Within such a vision, new and different accounts become an essential component of any imagining of change — let alone move towards it.

4.4 From remote modernity to intrinsic sustainability: A role for social and environmental accounting?

If we were to understand ecological accounting as only providing physical accounts of ecological systems then, as I have already mentioned, whilst I and other accountants/social scientists may have something to offer in assisting biological scientists and the like in constructing and communicating these accounts, this is not where our focus and efforts have lain to date. It makes sense, I suggest, to see if the emergent field of social and environmental accounting actually has something to contribute to our attempts to try and approach some notion of what ecological accounting is and/or might be.

The relatively short history of social and environmental accounting theory and practice (see, for example, Deegan, 2016 for a challenging review) has predominantly adopted a model derived from financial accounting — the process whereby organisations are held accountable to their financial stakeholders. On the whole, social accountants have sought to explore how organisations might produce information that provided a report of that organisation's interactions that was wider than simply its financial acts. The emphasis has tended to be on what data organisations could, should or actually did produce about such matters as their impact on human rights, employee conditions, community engagements, use of animals, influence on governmental processes, the safety and reliability of products and, most pertinently, the organisation's interactions with and impacts upon the natural environment. Whilst legislation exists governing the reporting of some of these issues, in general terms, this disclosure by organisations remains largely voluntary — and consequently very partial and predominantly self-serving. Certainly, no full and formal accountability is discharged by the current practices of organisational reporting we see worldwide (Moneva *et al.,* 2006; Gray, 2006; Milne *et al.,* 2009; Laine, 2010).

Now it might be argued that such accountability is unnecessary and that the purpose of social and environmental accounting is to help organisations make better and more sensitive choices — that we should judge social, environmental and, ultimately ecological, accounting by managerial criteria (Schaltegger *et al.,* 2017). Whilst it remains likely that such innovations in accounting are more likely to be adopted if they can demonstrate a benefit to the organisation itself, it is very far from clear that the organisation can ever put society and the environment ahead of its own interests: and this seems to be especially the case when the organisation is a large quoted company (Harris, 2013; Gray, 2006). Consequently, it is more typical to find proponents of social and environmental accounting arguing that organisations must be fully held to account and that a full social and environmental accounting is necessary to do this (Gray *et al.,* 1997, 2014).

Quite clearly, it has not been in the interests of most organisations to adopt — or support the adoption of — the mechanisms of full accountability. By contrast, organisations — and, in particular, large corporations — have used a variety of accounts to maintain an aura of illusion about their social responsibility and their environmental stewardship. The extent of this illusion becomes starker when organisations begin to make claims about their sustainability. A plethora of business-based reports have sought to encourage organisations to develop "full disclosure" about their own sustainability and, more particularly, there have been a series of serious major initiatives such as the Global Reporting Initiative (GRI) and Integrated Reporting which have provided support and sought to institutionalise the reporting of organisational relationships with sustainability and sustainable development (see Gray *et al.,* 2014 for more details). Despite this, it remains the case — certainly at the time of writing — that no organisation has come close to reporting on its sustainability — despite the enormous increase in corporate publications with titles such as "Report on Sustainability" which might lead one to suspect otherwise (see, for example, Buhr *et al.,* 2014). Basically, if one wanted to establish whether an organisation was (or more likely was not) sustainable (even in the Brundtland sense of the term) then it would be impossible to determine this from the reporting contained in sustainability (or, more accurately, "sustainababble") reports (Buhr *et al.,* 2014). The same story more or less obtains for (what might be thought of as) social sustainability as exemplified by the exhortations for organisations to report against their contributing to the United Nation's sustainable development goals (UN SDG) where, despite a great deal of fanfare, progress remains, at best, patchy (Global Footprint Network, 2017; Spangenberg, 2017; Bebbington and Unerman, forthcoming).

The continuing refusal of organisations to adopt substantive voluntary disclosure and the continuing refusal of governments to acknowledge that voluntary reporting is simply not working and shows no sign of ever doing so, has led to widespread dissatisfaction. In response, many researchers and pressure groups have chosen to abandon appeals to

```
• Accounts of capitalism
    – e.g. Collison et al. (2007, 2010)
• Accounts of un-sustainability
    – e.g. Gray. (2006)
• Accounts of the oppressed/silenced
    – e.g. Cooper et al. (2005)
• Accounts of the profession/corruption
    – e.g. Sikka. (2010a, 2010b)
• Silent + Shadow Accounts
    – Gibson et al. (2001); Dey et al. (2011)
• External Social Audits
    – e.g. Harte and Owen, (1987)
• Counter Accounts
    – e.g. Gallhofer et al. (2006); Steiner
      (2010)
• Performance-portrayal gaps
    – e.g. Adams (2004)
• Accounts for sustainability
    – e.g. Bebbington (2007)
• Accounts as imagining
    – e.g. Davison and Warren (2009)
• Accounts of Supply Chain + Accreditation
    – e.g. Locke et al. (2007)
• Regional Accounts of Water, Air, Land, etc.,
    – e.g. Lewis and Russell (2011)
Taken from Gray et al. (2014)
```

Figure 4.1.   Examples of Counter-accounts Emerging from Civil Society

self-reporting by organisations (see, for example, Medawar, 1976) and have chosen to seek to enforce an accountability directly. These exercises in "speaking truth to power" manifest themselves in a number of ways and are generally known as "external social audits" or "shadow-" or "counter-accounts" (see, for example, Thomson *et al.,* 2015; Gray *et al.,* 2014; Ch. 10; and see Figure 4.1). In essence these counter-accounts are based on the idea that if an organisation will not develop its own account-ability then other people will do it for them and the ensuing accounts — whether videos, documentaries, articles, reports or imitation corporate documents for example — challenge the organisations concerned and directly offer a questioning to institutional claims, (Adams, 2004).

Whilst the focus of the counter-accounts generally remains on organisations and institutions, this is by no means a necessary requirement and accounting scholars and activists such as Christine Cooper have opened up the ideas of a social accounting which addresses directly current social (and environmental) issues, (see, for example, Cooper *et al.*, 2005). Others have sought to widen the scope of counter-accounting (see, for example, Dey *et al.*, 2011) and have, as Laine and Vinnari (2017) say, sought to:

> challenge the usual norms of accounting, which place the organization in the focus of the account ... and could thereby enhance the visibility of humanity's socio-ecological interdependencies.

And it is this reasoning that has led to the slow but steady growth of (social) accountants looking to apply their expertise and insights into ecological systems directly (Russell *et al.*, 2017). As Russell *et al.* (2017) report we can now find the accounting literature paying attention to the accounts of bees, forests, rivers, lakes and blanket bog. These emerging initiatives of what are, more strictly, ecological accounts have yet (as far as I am aware) to reach critical mass and form a substantive presence in the social science literature. Consequently, I remain unsure where their impacts will be, how they might develop and what we might, at this stage, say about them. They represent an important strand to the future and one which should be watched closely. There is little more that I can say about them here.

4.5   Social and environmental accounting and sustainability: The story so far?

An ecological account of an organisation — at least an organisation of modernity and of late financial capitalism — makes little or no sense. The accounting entity will not be coterminous with ecological systems and will, the larger it is, intersect and influence a wide range of natural systems. If we are to attempt to expose, challenge and (perhaps) re-direct many of our economic and institutional entities then there is an urgent need for accounts that can achieve (some of) this in systematic and coherent ways. As Lehman (2017, p. 33) has it: we very much need an enabling accounting with regard to our ability to render our world intelligible and meaningful.

Whilst the mainstream of social and environmental accounting has generally maintained an anthropocentric approach to the determination of social responsibility and environmental stewardship, the advent of accounting for or about (un)sustainability has encouraged a planetary and societal focus: an outside in — rather than an inside-out — focus on the organisation. Measurements of global un-sustainability and projects such as the SDGs offer a broadly empirical basis against which to judge humanity's current ways of organising and, in particular, the claims of the organs of capitalism and modernity to be sustainable and just.

In *Accountability, Social Responsibility and Sustainability* we identify four ways in which the construction of an account of an entity's (un)sustainability might be approached.

Fully monetised accounts reflect the way in which a traditional economist might approach the issue and essentially comprise turning all of nature into a very particular form of exclusively human measurement. As a traditional economic mentality effectively excludes all of that which is natural — except insofar as it is manifest in prices (as does traditional financial accounting) — the approach is a consequence of the very reasoning which causes the alienation of humanity and nature: there is no way in which ecology can be present in economics except by reducing it to all that which it isn't. If, as economists tend to argue, the use of financialisation is the best way to motivate business (by speaking its language) to protect the planet, then perhaps this approach has a place here, (see, for example, Gaia and Jones, 2017). I would sense that it is the very antipathy of that which is needed for intrinsic sustainability to have a chance of success.

Integrated Accounts were very popular with business and politicians in the early years of the 21st century as they were claimed to be the Holy Grail of green and responsible business. The approach was exclusively organisation-centred and sought to identify flows of information around the social, environmental as well as the economic interactions of the entity and treat them as inter-dependent: showing how profit was derived from responsibility and from environmental stewardship. I remain totally bemused by these claims which seem to have nothing of sustainability or ecology within them and, at best, seem to be designed to show the less un-sustainable organisations in a more favourable light (see for example, Hopwood *et al.*, 2010).

Multiple Accounts essentially take the organisation and its activities for granted and encourage the production of three separate accounts governing the organisation's financial interactions (largely covered by financial accounts), its social engagements and its environmental influences/impacts. These three separate accounts, often known as the triple bottom line (TBL) offer, at best, three visions of the organisation and, if the social and environmental accounts are reliable and approaching complete, they will demonstrate the social and environmental costs that the pursuit of economic profit entails. Whilst I would argue that the approach has much to commend it, its principal weakness is also clear: the quality and completeness of the social and environmental accounts remains woeful whilst such reporting is voluntary. We see, in effect, organisations using the social and environmental accounts to demonstrate their beneficence — not to demonstrate their negative social and environmental impacts. Furthermore, for managers and those of a more modernist and calculative bent, the TBL offers no metric — no mechanical means — by which the social and environmental impacts can be formally assessed against the economic gains. Trade-off is elusive.

Sustainability Accounts are more ambitious accounts which seek to capture the degree to which an organisation or an institution has contributed to or detracted from (Brundtland style) sustainability. Few examples of substantive attempts in this direction exist in practice but researchers and activists have experimented with possible methods — typically within one of the three prior frameworks (*see also* Bebbington *et al.,* 2001). So, for example, Bebbington and Gray (2001) explore what it *would* have cost an organisation to operate if it *had* been "semi-strong" sustainable — the answer is considerably more than the economic profit for the period (Gray, 1992). Other attempts have tried to assemble indicators of local ecological health as the basis of their accounts (Ranganathan, 1998) or begun to develop mechanisms for measuring and accounting for biodiversity (Jones and Solomon, 2013; Gaia and Jones, 2017). Perhaps most promising, have been the attempts to derive the ecological footprint of the organisation, (Bebbington *et al.,* 2001). Equally, there are attempts to develop reporting against the UNSDGs (Bebbington and Unerman, forthcoming). What they all have in common is that they collate evidence which demonstrates that mankind's dominant ways of organising are clearly unsustainable.

It is a matter of some contention whether this effort directed towards the accountability and/or the unsustainability of organisations is worthwhile or not. On the one hand, it is pretty apparent that organisations are in nearly all cases manifestly unsustainable and that their very *raison d'etre* is in conflict with the essential needs of humanity and ecology. There remains, it would seem, a complete refusal on the part of mankind to acknowledge this and, perhaps, do anything about it. Would substantive accounts change anything? That is not clear. The central thrust of social and environmental remains, I would maintain, that we must push for full accountability (it is after a relatively cheap thing to undertake) and that until there is widespread evidence of the manifest un-sustainability of entities, politicians, business leaders and academics will continue to ignore the substantive issues and continue claiming that only a head in the sand approach is realistic. Accounts can, just maybe, stop this from continuing until it is so blatantly and devastatingly too late.

4.6 Reflections and conclusions?

This short essay has not sought to offer any particular utopian vision — although, maybe, it is the lack of accounts of possible utopian futures that inhibit our current imaginings. Certainly, it seems highly plausible that mankind urgently needs a variety of utopian imaginings to offer the prospect of something other than either the present catastrophe or the dystopian visions of the future that cloud literature and other media. Rather, the essay has been predicated upon the notion that modernity and, especially, its manifestations in finance, accounting, multinational business and neo-classical economics effectively suppress any substantive possibility of the emergence of utopian yearnings. In seeking to challenge the dominant narrative(s), ecological accounts — broadly understood — provide a crucial component in any hope that we have that mankind, its fellow species and its planet might have a future of any desirable sort.

Ecological accounting, understood as broadly as we have discussed here, is really in its infancy but, more crucially, is seriously resisted in all the places where it is most urgently needed — around the organs of modernity. Ecological accounts, as do social and environmental accounts, seek to "explore[d], examine[d] and critique[d] efforts to make (in)visible the impacts and interconnections of humans, their organisations and non-human worlds" (Russell *et al.*, 2017, p. 1428). The resistance in practice to such a self-evident good is bewildering. Such resistance requires some attempt at explanation. Perhaps we might follow Abbott's understanding that "in this electronic age we can be ignorant faster than ever before, about more things and in more settings" (Abbott, 2010, p. 174) and the distraction and focus elsewhere allows the major elephants (threatened with extinction though they might be) in the room to be ignored. Alternatively, we might understand such bizarre behaviour as a manifestation of what Alvesson and Spicer (2012) call "functional stupidity":

> Functional stupidity is organizationally supported lack of reflexivity, substantive reasoning, and justification. It entails a refusal to use intellectual resources outside a narrow and "safe" terrain. It can provide a sense of certainty that allows organizations to function smoothly. This can save the organization and its members from the frictions provoked by doubt and reflection. Functional stupidity contributes to maintaining and strengthening organizational order. It can also motivate people, help them to cultivate their careers, and subordinate them to socially acceptable forms of management and leadership. Such positive outcomes can further reinforce functional stupidity. (p. 1196)

Conventional accounting seems to be designed to reinforce functional stupidity when it comes to social, environmental and ecological matters. It supports and enhances the destructive power of multinational companies, of international financial markets and of the logic(sic) of neo-classical economics. If ever there was a need for counter-accounts, it is here. And whether such accounts are produced by the organisation or by stakeholders is really not the issue: the quality and reliability of the accounts is all that really matters if we are to stand any chance of getting out of this mess into which we have so gleefully driven ourselves.

It seems obvious, therefore, that all forms of ecological accounts must be explored and supported — subject only to the proviso that they be substantive, reliable, aimed towards completeness and driven by notions of ecological, planetary and social need. Our need for accounts of our planet; accounts of the effects that our major entities are having on that planet; and accounts of how our future might look are quite essential. They are also neither that difficult nor that expensive. Ultimately, as Lehman (2017) says:

> accounting and business research can assist in developing practices and structures which ……. create[s] a vision of commonalities between humanity and nature which connects with nested public spheres (such as grassroots environmental movements) to offer alternatives and new directions. (p. 35)

Why would mankind not choose to embrace such a vision? Currently, sadly, one can only speculate. Perhaps we just need more accounts of humankind's collective idiocy and less examples of mankind's (sic) heroic narratives.

Chapter 5

ECOLOGICAL CIVILISATION

5A — China's Emerging Ecological Civilisation

Liu Zhen[*,§], Mingyue Fan[†,¶] and John Margerison[‡,||]

Aizairenjian Consultant and Education Limited, Beijing, China
†*Jiangsu University, Jiangsu, China*
‡*De Montfort University, Leicester, UK*
§*janelz_liu@qq.com*
¶*fanmy520@163.com*
||*jpmargerison@dmu.ac.uk*

5A.1 Introduction

This chapter examines China's "ecological civilisation" policy using Sen's (1989) "capability" approach. It should be noted that there is a strong philosophical tradition in China based on the three ancient philosophies of Buddhism, Taoism and Confucianism. It has been argued (Pan, 2011) that they have created an ecological wisdom of the ages based around ideas that are still influential in China today: harmony between heaven, man and Nature (the earth); reciprocity; all under heaven (Tian

Xia in Chinese); and, an active self-regulating Nature. These ideas resonate with the details in the policy outlined below.

The contingencies in place that have led to this policy change in China include: the influence of ancient philosophy; embracing of new science; and, the disappearance of specialisation. The argument is that these contingencies are leading to episteme change (Foucault, 2002) in China towards ecological civilisation. At the same time this change to ecological thinking can be seen to accord with Sen's (1989) ideas; in that the Chinese people have been acknowledged to, in many cases, lack capabilities — for example: to be able to breathe clean air; drink pure water; work with fair working conditions; and, move and work wherever they wish.

5A.2   Ideas of change — Foucault's work on episteme change

There are at least six different theories that can help to explain why change takes place in society: Realism (Bhaskar, 1978); Paradigm Shift (Kuhn, 1962); Structuration Theory (Giddens, 1991); Post-modernism (Lyotard, 1993); Episteme change (Foucault, 2002); and, Modes of Existence (Latour, 2013). The episteme change theory of Foucault is used in this chapter since it provides a detailed analysis of "Modern" which enables a meaningful comparison with Chinese thought.

Foucault (2002, p. 183) explained an *episteme* as: "In any given culture and at any given moment, there is always one *episteme* that defines the conditions of possibility of all knowledge, whether expressed in a theory or silently invested in a practice." Birkin and Polesie (2011) interpreted Foucault's notion of an *episteme* as the possibility of knowledge — what knowledge makes possible — the consequences of which define an age. Foucault (2002) was able to identify different epistemes and what transformations could be attributed to episteme change. He noted that a new possibility of knowledge undermined the credibility of the existing episteme (*ibid.*).

Based on Foucault's (2002) theory of episteme change, Birkin and Polesie (2012) argued that there is now a new possibility of knowledge, of episteme change, from Modern to what they called Primal. The

Modern episteme (since 1800) was based around abstract, anthropocentric, logical belief systems and this, they have argued *(ibid.)*, has led to unsustainable development. The emerging "Primal episteme" is based on the knowledge provided by scientific studies *(ibid.)*. Hence at the centre of a primal episteme is empirically grounded science such as thermodynamic dissipative structures (Hammond, 2004), with accompanying developments in mathematics such as chaos theory (Gleick, 1988). This in turn is a new ontology which finds support in China in ancient ideas and philosophies (Birkin and Polesie, 2012). In this chapter it is argued that aspects of China's traditions, cultures and beliefs resonate well with the new science and with the Primal episteme/ ecological civilisation.

5A.3 China's ecological civilisation policy explained

Senior Chinese government figures (Pan, 2011) from 2007 outlined the concept of ecological civilisation (sometimes called eco-civilisation) with harmonious society being incorporated into the rhetoric. Oswald (2014) noted that in 2007 at the 17th National Congress of the Communist Party of China, Party General Secretary Hu Jintao announced a new model of growth incorporating "ecological civilisation" to replace the old unsustainable industrial model "industrial civilisation".

It has been noted that whilst it created unprecedented levels of material wealth, the industrial model of development, based on high levels of resources and energy consumption, also brought serious pollution and ecological destruction to the industrialised world (Ma, 2007). It was further noted that global capitalism had transferred the most polluting, resource-intensive and high-risk manufacturing industries to developing countries *(ibid.)*. This had allowed developed countries to alleviate the pressure on their own environments without making any changes to their model of growth *(ibid.)*. In this context China developed its industrial economy at the expense of heavy environmental degradation which was seen to be unsustainable *(ibid.)*. Hence the rejection of modernity is backed by serious ecological damage both to humans and Nature.

Ecological civilisation was enshrined in the 12th five-year plan (2011–2015). In 2013 the official organ of the Central Committee of the

Communist Party of China (CCCPC) outlined the following basic features of ecological civilisation (Jiang, 2013):

> "First, human beings are a part of nature. The relationship between human beings and other creatures should be one of equality, friendship, and mutual reliance, as opposed to a relationship in which humans are supreme.

> Second, since it is nature that has given us life, we should feel gratitude towards nature, repay nature, and treat nature well. We should not forget the debt that we owe to nature, or treat nature and other creatures violently.

> Third, humans are entitled to exploit natural resources, but we must take the tolerance of ecosystems and the environment into account when doing so in order to avoid over-exploitation.

> Fourth, human beings must follow the moral principles of ensuring equity between people, between countries and between generations in resource exploitation. We should refrain from violating the rights and interests of other people, other countries, and future generations.

> Fifth, we should advocate conservation, efficiency, and recycling in the utilization of resources so as to maximize efficiency whilst keeping consumption and the impact on nature to a minimum.

> Sixth, we should view sustainable development as our highest goal, rejecting the overexploitation of resources and short-sighted acts aimed at gaining quick results.

> Seventh, the fruits of development must be enjoyed by all members of society and not monopolized by a small minority."

5A.4 Chinese ancient philosophical traditions

In this section, firstly; the modern interpretations of the ancient traditions (focusing on Buddhism, Taoism and Confucianism) are discussed in terms of their links to sustainability (mainly environmental sustainability) and secondly; recent research looking at Chinese accountants' environmental ethic is used to argue that the traditions and their modern interpretations are still relevant in modern China. The purpose is to add empirical findings in support of the episteme change to ecological civilisation taking place in China now. If this is the case, then it can be posited that changing episteme together with accountants' imbued with ecological thinking will lead to changing accountability for sustainability. This changing accountability is evidenced later in the chapter.

Tu (1998) has argued for Confucian values to replace the rampant self-interest of the modern era via a return to ideas of harmony between humans, the earth and heaven. Taoist philosophy acknowledges the importance of a balance between yin and yang and hence the idea of harmony similar to that in Confucian teachings (Palmer and Finlay, 2003). Buddhist thinkers have rejected the modern hierarchical dominance of humans over nature and emphasised the need for a transformation of values and lifestyles if the ecological crisis is to be overcome (Swearer, 1998). Pan (2011) has argued that with *Tian-xia* thinking that fuses the three ancient philosophies the concept of the harmonious unity of man and Nature is central.

In Foucauldian terms a new episteme emerges when the viability of the old episteme is undermined and the contingencies in place support a change in episteme. In terms of viability of the modern/industrial episteme in China there is evidence from the textual sources above that there have been viability problems in China. There is also evidence that an emerging ecological civilisation will overcome these viability problems. In particular: (1) right and proper conduct — a morality based around harmony and reciprocity with a deep-seated reverence of nature going back thousands of years; (2) specialisation — a move away from the modern dominance of economics towards a wider appreciation of the need for all disciplines to work together in harmony to solve sustainability problems (a capability approach); and (3) anthropologization — moves away from thinking that humans are seen to be in control of the planet and its species towards an anthropocosmic world view where humans are part of a greater cosmos or universe (Tu, 2001) and so more holistic solutions are forthcoming.

Further, the textual sources support the notion of China having special characteristics that make the possibility of episteme change more likely:

(1) The emergence of a new metaphysics based on a life-centred morality — around modern interpretations of ancient philosophy,

with a strong emphasis on harmony between humans and nature (Miller, 2003; Pan, 2011; Sponsel and Natadecha-Sponsel, 2003);

(2) Invention linked to new science — the notion of scientific development is central to Communist Party of China (CPC) ideas of ecological civilisation and includes the embracing of new technologies and the move to a low carbon society, with China at the centre of invention and new science as evidenced in the literature (Chan, 2014);

(3) The disappearance of specialisation; and, (4) a more integrated view of the universe — the way in which China and its government have encouraged holistic, whole society moves towards change — with a cross-disciplinary approach — as opposed to the economics-based, over-specialized approaches of the Modern episteme (Ma, 2007). At local level this has been evidenced in the report by Liaoning Environment Protection Bureau (EPB) (2008) with the ideas of a scientific concept of development and the construction of an ecological province. This demonstrated the fusing of many disciplines and a whole society initiative to preserve nature. These EPB reports are analysed further in the next section on accountability.

Chinese accountants' environmental ethic was tested by surveying three groups — Chinese accountants; Chinese accounting academics; and, Chinese undergraduate accounting students (Table 5A.1). The surveys were carried out using judgemental sampling based on databases held by the

Table 5A.1. Summary of Responses to Survey Question on Accountants' Eenvironmental Ethics

Statements About Your Attitude Towards the Natural Environment	% Answering Strongly Agree or Agree	Mean Score
(R1) Humans should be one with the natural environment (Buddhist philosophy) (Sponsel and Natadecha-Sponsel, 2003)	91	1.7
(R2) There should be harmony between heaven, earth and humans (Taoist philosophy) (Miller, 2003)	83	1.8
(R3) Human flourishing can only take place within the larger matrix of nature (Confucian philosophy) (Berthrong, 2003)	59	2.3

authors. In total over 100 responses were obtained and it was the questions on the ancient philosophies that were particularly interesting in this context:

These findings certainly accord with the hypothesis that in modern China a group with influence (now or in the future) over accountability matters (accountants) appear to be influenced by the ancient philosophies. A proportion of respondents also were influenced by more than one philosophical tradition, even all three, and this gives credence to the idea of the three traditions flowing into one (Clayre, 1976).

5A.5 New forms of accountability in China

The way in which organisations in China have changed and developed their accountability for sustainability over the last 10 years, we argue, provides evidence of episteme change to ecological civilisation. This in turn can be seen as an expansion of capability for the Chinese population — workers, consumers, communities around firms — with better, happier and healthier lives. In this chapter, the link between Chinese accountants and the accountability changes cannot be clearly shown, but if Chinese accountants can be thought to represent educated Chinese citizens then it is the larger group that must be influential in the accountability initiatives.

To examine this changing accountability the analysis is split into two parts — external and internal accountability. "Accountability" is used to describe all techniques, including accounting and reporting in both calculative and narrative forms, by which an organisation gives an account of its activities to those people and groups that have a right to that information (both internal and external) (Gray *et al.*, 1997; Kamuf, 2007; Roberts and Scapens, 1985; Staubus, 2003).

5A.5.1 *External accountability by companies and government*

Typically external accountability information in the accounting for sustainability area has comprised both calculative and narrative (Kamuf, 2007). Calculative information is interpreted to mean information using accounting numbers including techniques, such as Full Cost Accounting (Bebbington *et al.*, 2001) and Ecological Foot-Printing

(Wackernagel and Rees, 1996); whilst narrative information is based on words, such as sustainability reporting on environmental aspects and impacts and the measures in place to manage the environmental aspects or on fair treatment of workers both in the organisation and in its supply chain. The Global Reporting Initiative (GRI, 2013) has provided guidance on the sorts of disclosures to be made by organisations in sustainability reports that have been widely used around the world.

There is empirical evidence of the growing use of environmental accounting and reporting by larger Chinese companies (Gao, 2009). Gao's research and other surveys (Guo *et al.*, 2008) showed a growth of CSR disclosures by large Chinese companies in the period since 1995. This was borne out by the authors' visit to and analysis of disclosures by a major Chinese steel producer, where all the sustainability reporting (in both Chinese and English) was similar to best practice by Western companies.

Chinese research on disclosure by Chinese companies of environmental accounting information (Zhang, 2010) has presented a pessimistic view that there is a lack of concern by government and the general public about such disclosure. The suggestion was made that investors tended to invest in any business which would bring high profit and tended to ignore environmental effects. Zhang recommended that: the idea of Green investment should be advocated; naming and shaming of heavy polluters; a national database of environmental reports of companies; and, pressure on companies to publish environmental accounting information voluntarily (*ibid.*).

Recent research (Du and Gray, 2013) on the emergence of stand-alone social and environmental reporting in Mainland China begged the question as to what were the drivers for entities to disclose environmental information and suggested that there is much to be done on this question. Xiao (2006) suggested that environmental accounting and reporting in China was needed so that companies could integrate environmental issues into the business agenda and discharge their environmental accountability. Hence the best current explanation for external accountability by major quoted Chinese companies is around institutional theory (Larrinaga-Gonzalez, 2007) and in particular coercive pressures from government and mimetic pressures from major international competitors.

But it is the absence of reporting in most cases below the major quoted company level that is worthy of reflection and further research. Choudhury's (1988) paper on absence in accounting suggested that one reason for absence is that it implied trust. This suggests that when an organisation is in a position of trust by society it will not see it as necessary to account and report. In sustainability accounting and reporting terms perhaps the absence of trust is what motivates Western companies and their big Chinese counterparts to account and report on environmental matters. Perhaps it is the trust by stakeholders in the majority of Chinese companies that leads to the absence of external reporting. Du and Gray (2013) noted suggestions of a negative relationship between state ownership and disclosure but their own findings did not bear this out. The implication of this negative relationship is that when the state owns an enterprise the key reporting will be to the Ministry of Environmental Protection (MEP) — accountability to the key stakeholder — the State.

A key empirical contribution in this chapter is to extend the analysis of sustainability accounting and reporting (in particular environmental sustainability) to political regions that are based on Chinese provinces and Provincial Environmental Protection Agencies (PEPA). There are 31 provinces and autonomous regions (AR) in mainland China. In 2015 a review of all these provinces and ARs was carried out and it was found that they all produced an annual report on environmental protection and in most cases had been doing so for many years. These reports are freely available on the web sites of the relevant PEPAs. The earliest reports were for 1995 and all had produced a report for 2014. These reports were in Chinese with no English translations provided.

For the purposes of this chapter, one province — Liaoning — was selected and its recent reports analysed around environmental themes and some of the trends of language used and statistics analysed. Liaoning was selected because this research included interviews with two officials of Liaoning PEPA during a visit to that province in 2012. The Liaoning reports from the five years to 2014 were translated by the Chinese member of our research team. In translation the Liaoning reports have been titled: "Environmental Situation of Liaoning Province" and were prepared by the PEPA. Of particular interest in the context of this chapter's focus on ecological civilisation; for each of 2013 and 2014 the report was subtitled: "Strengthen the construction of ecological civilisation". Also the 2013 report specifically referred to the fact that ecological civilisation reform had started, leading to the setting up of an ecological province.

For all the reports from 2010 to 2014 statistics were provided on the following areas: ambient air; precipitation; urban centralized drinking water source areas; reservoirs; offshore sea area; noise pollution of road traffic; ecosystems; and, emissions status. The range of statistics provided was very wide and for the purposes of this paper the important thing to note is that they represented the results for the whole province — state, companies and domestic households. There was no indication as to how the statistics were gathered and there was certainly no assurance or audit of the figures by an independent party.

In terms of environmental accountability the following matters were disclosed together with performance against a 2010 base: assessing the number of cities and towns with nitrogen dioxide conforming to an MEP standard; acid rain frequency in cities; water quality in reservoirs and in offshore sea area; decibel levels of road traffic; percentage of the province area of excellent; good or standard ecological quality; and, carbon-dioxide emissions in tonnes and reductions each year. As a whole the package of disclosures it shows a province taking a very close look at its environmental aspects — a sense of provincial ecological responsibility. The wide range of activities being reported on give the impression of a provincial EPA that seeks to be accountable for all the activities in the province and has moved towards this in the depth of its disclosures.

Interviews were carried out with officials in two local offices of the Liaoning PEPA. They outlined the systems for monitoring and, where appropriate, reprimanding and penalizing companies that were operating outside acceptable limits. A very interesting point was raised by one official about shared responsibility; this was in relation to a pollution incident involving China Petroleum Company at Dalian petroleum pipelines where the company shared responsibility with the MEP. This suggested a more cooperative approach than would typically be the case in the West.

5A.5.2 *Internal accountability — companies and environmental management accounting*

Sustainability in a Chinese ecological civilisation setting has tended to concentrate on environmental sustainability. In internal accountability this is often classed as environmental management accounting (EMA). A definition of EMA is given by the United National Expert Working Group (IFAC, 2005, p. 19):

"EMA is broadly defined to be the identification, collection, analysis and use of two types of information for internal decision making:

- Physical information on the use flows and destinations of energy, water and materials (including waste) and
- Monetary information on environment-related cost, earnings and savings."

Part of the research on which this chapter is based involved (1) interviewing accountants at 20 Chinese companies in 2012 and (2) two case studies of Chinese Tourism Companies X and Y:

5A.5.3 *Interviews*

A number of environmental management accounting initiatives were described by the interviewees:

- information on costs involved with waste water treatment;
- measurement of pollution related costs of new factories in business planning;
- identification of costs of equipment recycling;
- estimates of remediation costs associated with incineration plants with associated liabilities;
- accounting for bottle recycling projects;
- measurement of energy reduction initiatives;
- detailed data generated on energy usage and pollutants;
- accounting for planned introduction of thermal pumping and associated reduction in coal burned annually;

- costing of tree planting programmes;
- costing of energy saving and emissions treatment programmes;
- costing of dust control measures;
- measuring carbon dioxide emissions of the company.

It can be concluded that, internally, the Chinese companies discussed by the interviewees had a highly developed sense of environmental impacts and were developing new accountings to account for these impacts.

5A.5.4 *Case studies*

Two case studies of Chinese tourism companies led to the gathering of data on EMA and to reveal characteristics of intrinsic sustainability in China, as well as how and why people's behaviour changes as time passes. The cases show evidence that modern interpretations of Chinese ancient traditions have heavily influenced individuals' consciousness, knowledge and behaviours towards being sustainable. Employees in the two tourism companies were interviewed and studied between 2014 and 2015.

The first case study — Tourism Company X — is a state-owned company located in an ancient town in Zhejiang Province. The province is one of the most wealthy and well-developed industrial areas in China. The strong power from local residents (mainly from one family stretching back two thousand years) has constrained the changes towards EMA by the company. The company is supported by the local government (one of the key local officials is the leader of the company). The power dynamic between local culture/residents and the government leant towards the former. As a result, change was not happening and there was no evidence of any environmental projects and implementation of EMA tools during the study. Because the strong character of local society as sampled is very unique and attractive, it also maintains strong self-regulation that becomes a constraint to new things including EMA implementation. Without this strong self-regulation, this current structure would have collapsed many years ago.

It seemed that local society had strong power due to its ancient links. It was in favour of conserving and retaining the ancient town in modern China. As a result it was the strong barrier to any changes. This special context had provided the concept of "father authority" in families traditionally, that did not only empower individual families, but also empowered the whole society through the connection of "blood". The strong power had been able to block change coming from outside.

However, the power from the local society, supported by traditional culture, contributed to the avoidance of the exploitation of natural resources and the preservation of historical cultural sites. It is important to recognise that the aim of Ecological Civilisation (EC) is not only to focus on protecting the natural environment, but also seeks to improve the capability of people — living a good and healthy life — according to the 12th five-year plan announcement in 2013 the official organ of CCCPC (Jiang, 2013).

So, in the first case study changes in internal accountability of the company were not shown in terms of EMA implementation. However the power from the local society's authority was apparently perceived and accepted by people who lived or worked there. The respect for Nature and the environmental were therefore inherited.

State-owned Tourism Company Y is located at Guangdong Province in a coastal location. Tourism Company Y manifested the opposite direction of power influence. The "authority" was to accept new knowledge of being sustainable and enjoying the benefits from doing good, which gave the company an opportunity to shift their strategy towards sustainability. To this end it had achieved a significant outcome — internal EMA had been implemented.

It had initiated five environmental projects, lasting a decade, that had significantly improved its environmental performance and advanced stakeholders' quality of life. Although EMA tools were not applied consciously and systematically from the beginning, they were gradually set up. While internal accountability was developed, becoming more and more visible and measurable, it substantially changed people's knowledge from the top down, and helped managers to make better decision.

In Tourism Company Y, individuals or authorities (individuals or groups) represent human agency. Their understanding and interpretation of the reality has an impact on the structure. Hence a series of environmental improvement actions were initiated by a small group on the project — Energy Saving and Emissions Reduction (ESER). The decision for ESER came from top management to staff. The power could be seen from the needs of the company and how to interpret the needs by the authorities of the company. The social structure changed and the pattern of individual's behaviours changed to a new dialectic of control, which led to a different outcome from the first case. As a result, a series of innovative environmental initiatives took place and the routine of individual's and group's behaviours changed as the project proceeded. ESER's resources were later expanded and strengthened to successfully advance further actions during change time (when the project was operating). Meanwhile new values, ideas, and customs were built up. The internal accountability was set up and improved accordingly.

This study found that there were four years from 2006 to 2009 from the setting up of ESER when little happened until 2010 when four big innovative projects were started. The benefits of improved environmental performance were evident from 2007. However, it did not influence people's changing behaviours towards being environmentally friendly until the company started the real innovation environmental projects in 2010.

The surrounding societies in the second case lacked the strong familial roots due to the town being new and populated mainly by immigrants from other parts of China. Although people would tend to trust the "authority" from familial relations, without complex relations and historical roots, the power of the "father" authority was weaker than the authority in the company. Instead the "manager" in the company had taken the role as "father" in this society.

Although the outcome was very different in each of the two cases, the pattern of power shifting was the same. It was evident that the relationship between people and the environment had strongly affected internal accountability. It can be concluded that "trust" in Guanxi between people and the environment (society), which is empowered by "the authority", is the key to engage with or refuse the direction of any change. The power is also influential in how people would understand the issue of sustainability or Ecological Civilisation, and receive this knowledge. In fact, the two cases were both in accordance with Chinese traditional thinking discussed above.

The evidence from the interviews and case studies strongly supports the notion that Chinese companies have been much more concerned with internal accountability for sustainability than previously thought. Most important is that the power of ancient traditions has been a key motivator of change in Chinese companies and government towards ecological civilisation.

5A.6 China's emerging ecological civilisation and the capability approach

As discussed in the introduction, episteme change in a Chinese context means a change in people to ecological thinking. This can be seen to accord with Sen's (1989) ideas; in that the Chinese people have been acknowledged to, in many cases, lack capabilities — for example: to be able to breathe clean air; drink pure water; work with fair working conditions; and, move and work where ever they wish. Hence the lack of capability drives change towards greater capability.

In Sen's terms (1989; Tennessee, ND) the basic features of ecological civilisation can be considered as a set of "functionings" leading to "capability expansion" that can be measured in outputs that are not traditional economic metrics. Thus, for example: all people have rights and interests that should not be violated — particularly succeeding generations; and, Nature should be respected so that it is not over-exploited (here Nature is seen as person in Sen's analysis). So, in ecological civilisation terms, humans have a key functioning — being able to live harmoniously with Nature so that they can breathe, consume, work and therefore achieve the capability of living a good and healthy life. This move away from purely economic metrics, that drove China both before and particularly after Deng Xiaoping's reforms (Huenemann, 2013), can be interpreted in Sen's analysis as a move towards a capability approach.

5B — Can Vietnamese Traditional Values Drive 'Sustainability'? An *Emic* Perspective

Lien Le Monkhouse

Sheffield University Management School, Sheffield, UK
l.l.monkhouse@sheffield.ac.uk

In this piece of work, I will try something I have never done before: write an emic account of Vietnamese culture and its impact on our relationship with Nature. An emic perspective, meaning the perspective of a person living within a culture, is one common approach used in anthropological and cultural research (Helfrich, 1999). The validity of the method rests on the premise that a person living within and having enculturated into their native culture has an insider's knowledge and tends to understand their own way of living the best. However, the work will also draw on research by other Vietnamese and foreign authors. I must also add that my recent experience living abroad has helped me better understand and appreciate my original culture.

I was born and brought up in Vietnam, a tropical country in South East Asia with the land covering a mere area of approximately 127,000 sq m (330,000 km^2) but is home to a large population of over 96 million (CIA, 2017). As with the rest of generations of Vietnamese schoolchildren, I was taught about our pride to have a "beautiful country with golden forest, silver sea and rich land" — a phrase having been imprinted so deep on my mind that I cannot forget after years of living and working away from my home country. I often wonder to myself how much longer will Vietnamese people remember that phrase, or indeed, how many more years will the future generation still enjoy these natural assets. Since declaring the nation's independence after the Second World War (Truong *et al.,* 2001), a deprived and torn-apart Vietnam has been in desperate need to use resources for rehabilitation and economic reformation. What a peculiar choice of country for the "sustainability" topic when it is doubtful whether the country has long passed the survival mode! Even more intriguing, the attempt here involves looking *backward* into Vietnamese traditions to evaluate the country's capability to move *forward* to embrace a sustainable way of life. By any means it is not obscure to look at traditional values, simply because culture, as well as values, ultimately is the way we live, that has been learnt and shared, and passed on through generations in a society (Linton, 1945).

With that in mind, I embarked upon a journey to search into my origin, to examine whether and how the way Vietnamese people live, our values, traditions and beliefs can drive "sustainability" — the mode of living that could turn around and save our planet earth from an alarming depletion rate because of human impact (Barbier and Burgess, 2017). Among a wide spectrum of factors that can lead to sustainability, I have chosen to discuss the pro-environmental aspect, i.e. sustaining the earth by living within its means, keeping a balance and not harming the environment. This piece will consider Vietnamese people's relationship with nature and how or whether that has been a part of our traditional values which can naturally push forward pro-environmental behaviour, and thus a more sustainable way of life.

5B.1 Vietnam — a distinctive cultural identity?

The word "cultural identity" in Vietnamese is "*bản sắc văn hóa*", with the word "*bản sắc*" as the literal translation of "true/aboriginal colour". So the question is whether Vietnamese culture has a different, unique "colour" (hence the rationale for this research) compared to, for example, China — a major civilisation and much more widely researched Eastern culture in the literature available to the West. Undeniably, throughout history, both Vietnamese culture and people have been known to be heavily influenced by the mighty northern neighbour China (Nguyen, 2008).

In fact, many Westerners cannot tell the difference between Vietnamese people and other East Asian countries from the look. Back at home, I was only asked if I took after my mother, or father, or anyone else in the (extended) family. I had never wondered if my look is of a typical Vietnamese until the day I started living abroad. Classmates, acquaintances, even passers-by, and later on, colleagues when first meeting me, still ask me till this date the same old question "Are you Chinese?" or "Are you Japanese" (and sometimes Thai, etc.). When I went on a research trip across South East Asia in the summer of 2017, a Chinese medicine doctor in Singapore insisted that I had "*Han*" (a major Chinese ethnic group) blood in me, just by looking at my toe nails as he treated a muscle strain on my foot, which I had suffered from an accident whilst playing folk games with Vietnamese children in Hanoi earlier in the trip. A couple of months later, a Vietnamese professor, during her visit to the university where I work, made a joke at my "farmers' feet" saying they were clearly of descendants of "*Giao Chỉ*" (ancient Vietnamese people). In fact this anthropological feature was noted in 1868 by *le docteur* Clovis Thorel (Simoën, 2013) in the French expedition of Doudart de Lagree as a distinctive feature of "An Nam" (i.e. Vietnamese) people. I suspect that both of these observations were rather myths than science, but certainly my identity has been less than obvious to Westerners' eyes.

A better-known personal narrative about Vietnamese people is regarding war stories (Roper, 2007). Among Vietnamese children born after the war, much of the history lessons I was taught at school was about the wars: over a thousand years under Chinese domination, ongoing resistances with the Chinese through early history, wars with Khmers, the Chams and the Mongols, then a long period of French colonialism (1858–1945), followed by the Indochina war against the French (1945–1954) then finally the American war (1954–1975) (Truong *et al.*, 2001). Being a small country lying at a strategic geographical location as the farthest south-eastern point of the Indochina Peninsula might have contributed to that unfortunate fate of Vietnam. The long history in wars has built up traditions, becoming a part of what makes Vietnamese people the way we are.

At the dawn of the Christian Era, Vietnam was a key port of call on the sea route for seamen, traders, priests from India before entering China or Japan, who took advantage of the South-eastern monsoon when setting out, and of the North-eastern monsoon the next year on their way back (Nguyen, 2008). Vietnam was subject to natural interaction with the region due to its geographical location at the cross-roads of great ancient civilisations such as India and China. In addition to this natural route of cultural exchange, Chinese dynasties also attempted to assimilate the Vietnamese nation during the thousand years of occupation in this country from 111 BC (Phan *et al.,* 1991). Apparently, Vietnam has been particularly and heavily influenced by China's political, social and cultural institutions. Vietnamese people have certainly adopted and adapted values, beliefs and philosophical traditions from other civilisations, and together with our native culture and own history, Vietnam has emerged with our own distinctive identity (Dao, 1939). In short, it has been recognised among Vietnamese historians that there are three layers of Vietnamese cultural establishment (Tran, 1999):

(1) Vietnamese indigenous culture from pre-historic time and *Văn Lang* — *Âu Lạc* (the first states of Vietnam);
(2) Cultural exchange with China and the region (predominantly India) during Chinese occupation of Vietnam;
(3) Cultural exchange with the West during French Colonial times and in modern history.

This three-layer framework provides a structure that helps explore Vietnamese people's beliefs, religions and traditions, and it will be referred to again later on in this work.

5B.2 Brief background of Vietnamese people's relationship with nature

To understand the religious and philosophical underpinning of the man–nature relationship in Vietnam, it is worthwhile to first understand the background context of how Vietnamese people live in the natural environment here. Ultimately, cultural heritage of a nation stems from their history of living in the natural and social environment, with the *natural* environment existing first *before* people and *outside* of their wills and creation (Dao, 1957). I still remember vividly the definition of the term "culture" taught across universities in Vietnam — well, our frequently used method of rote learning in the past could be subject to criticism, but that is exactly how I can recall word-by-word important pieces of knowledge taught decades ago, which can be roughly translated as:

> Culture is an organic system of material and spiritual values that have been created, accumulated by people through the process of practical life activities, during the interaction with the natural environment and social environment. (Tran, 1999, p. 10).

Vietnam is a peninsula that is geographically located in South East Asia region, where great rivers stream down from the Himalayas and Tian Shan mountain ranges. The downstream of these rivers, including the two running across Vietnam namely Red River in the North ("*Sông Hồng*" in Vietnamese) and Mekong River ("*Sông Cửu Long*") in the South, are naturally enriched delta regions. Twined with the hot and humid climate, monsoon and a huge volume of rainfall, these were the foundation for the birth and establishment of water-rice agriculture in Vietnam. Vietnamese history has recorded the existence of this dominant occupation from the early civilisations of *Hòa Bình* and *Bắc Sơn* from the Stone Age (Ha, 1998). Vietnamese people from early days lived based on water-rice agriculture, therefore there was the need of settlement in river valleys for people to wait till their plants turned into harvestable crops. Vietnamese agriculturists constantly needed to be on close watch of all natural forces such as the rain, wind, seasons, sun, etc. This means inevitably their relationship with Nature was at a different level of respect and long-term dependency than that of, for example, nomads from dry and cold climate in the North-Western part of the world, who travelled the land to find grass for their cattle (Tran *et al.,* 1998).

Vietnamese people accumulated wisdoms of the connection between various natural factors (for example what predicted weather condition and success of a harvest, etc.) and passed on these idioms of their under-standing of nature forces generation after generation. During my child-hood in the 1980s, when I was sent to live with my grandparents in Central

Vietnam in the summer holiday, I witnessed every morning my grandma would first use a pole to push up a huge wooden shutter for the light to come in the house (there were no locks back in those days in rural areas), and open the door to go outside for her weather check. She would look at the clouds on the horizon and their colours, feel the moist in the wind, perhaps watch the behaviour of certain type of birds in the garden, then tell me whether to bring a hat or a raincoat going out on the day. Quite effective I must say, considering we did not have an advanced scientific system of weather forecast at the time, especially in the countryside!

The climate and natural conditions of the country have shaped a history of understanding, respect and gratitude of Nature by Vietnamese people. It can be said that Vietnam has been favoured by nature for a rich flora (and fauna to some extent), with Mekong delta in the South dubbed as a "biological treasure trove" by international journalists (Fantz, 2008). However, it is by no means an easy relationship — the nation has a long history of on-going struggles against natural disasters, majorly hurricanes, heavy flooding, and draughts. According to the World Disasters Report 2015 by the International Federation of Red Cross and Red Crescent Societies (IFRC, 2017), Vietnam is one of the five countries most affected by natural disasters and climate change, with an average of 10–15 floods each year, impacting the lives of millions of people. Depicting the same level of devastation, yet images of torrential rains, water raising to house roofs, people flocking up into higher land to wait for aids and for the floods to withdraw, etc. are of stark contrast to those of cracked fields, withering and burning out crops, dying cattle lying around helplessly, etc. in the draught season.

On the other way round, the natural environment too has been a victim of destruction in history. During the Vietnam War (1954–1975), the US military used more than 14 million tons of explosives, on this small country alone more than twice the amount they had used during World War II, sprayed millions of gallons of defoliants that killed or burnt crops, forests and other vegetation, turning much of pre-war lush and green countryside, fertile river valleys into barren land (Vietnam War Reference Library, 2001). The destruction also came from Vietnamese people themselves, notably via deforestation to plant crops, build facilities or produce hydroelectricity (or simply to sell timbers!) (Pham, 2017). The reasons for this act against nature can range from means of survival, "sacrifice" for economic development (as with many other developing countries), to sadly a mere lack of knowledge and long-term visions.

The relationship between Vietnamese people and Nature has gone through the course of thousands of years of history, the same length of human existence in the country. What about the future? Do Vietnamese people possess the capability to further nurture this relationship in a bid to work towards a sustainable way of life? The answer may be revealed through an examination of Vietnamese spiritual life and cultural values, in the next part.

5B.3   Religious and philosophical underpinning of the man–nature relationship

After the war ended in 1975, the government of united Vietnam officially declared the state atheism while allowing Vietnamese people to have religious practice under the constitution (Lewis *et al.*, 2009). Though a large number of the population do not have formal membership in religious communities, Vietnamese people often practise and/or have reference to these principles in daily life (General Statistics Office, 2017). The Party-State of Vietnam has also identified selected religious beliefs, customs and rituals as part of national culture (Communist Party of Vietnam, 1998).

With regards to "traditional" values, the first two layers in the framework introduced earlier in this piece are brought forward for further discussion: first, Vietnamese original culture, and second, cultural exchange with foreign civilisations in early history. The indigenous beliefs held by Vietnamese native people had existed first, before the so-called "Three doctrines" (*Tam giáo* in Vietnamese), namely Buddhism (*Phật giáo*), Confucianism (*Nho giáo*), and Taoism (*Đạo giáo*) were brought to Vietnam under the influence of ancient China and India. However, among these three adopted traditions, Confucianism is omitted from this discussion of the relationship between man and nature. The reason is because Confucianism served the function of social and political structure and reform, hence mainly concerning moral codes and how to handle social relationships (between people and people) (Monkhouse *et al.*, 2013). It was succinctly summarised by Harvard Sino-Vietnamese historian

Hue-Tam Ho Tai when discussing Vietnamese religions, that "scholar-officials gave more weight to Confucian teachings; common people put more emphasis on Buddhism and on Taoism in its popular religious form" (Hue-Tam, 2017).

The third in the earlier-mentioned three-layer framework of Vietnamese cultural establishment is regarding cultural exchange in modern history. In the 17th century, Spanish, Portuguese and French missionaries introduced Catholicism into Vietnam, forming the largest among religious minorities in the country (Truong *et al.*, 2001). Then during the early 20th century, the French colonial rule officialised the Latin-based writing system that had been introduced a few centuries earlier by French missionary Alexandre de Rhodes versus the former ideographic system similar to Chinese characters (Truong, 2016). These are some examples of cultural elements absorbed from the outside world in Vietnamese modern history. They have less relevance on the discussant issue of this work, and "modern times" by nature is not what we consider as "traditions". Moreover, Western readers who have knowledge of Vietnam War (1954–1975) may also raise the question of why discounting the impact of Marxist-Leninist ideology in this analysis. Though the Vietnamese fought the war for national liberation in the name of communism, this orthodox rhetoric was majorly used for political and administrative discourse (Palmujoki, 2016); it did not permeate the wider society in a meaningful way to impact on daily decision-makings. As a result of this exclusion approach, the most relevant religious and philosophical systems that are chosen next to further the discussion comprise of Vietnamese folk beliefs and indigenous culture, Buddhism, and Taoism.

5B.3.1 *Vietnamese folk beliefs and indigenous culture*

As the term suggests, this part of Vietnamese culture has been passed on to contemporary society through myths, legends, folk stories and songs. Arisen from ancient time, Vietnamese folk beliefs (*"tín ngưỡng dân gian"*) consist of beliefs in our origin, in fertility, and worship of nature and of man. From early lessons at primary school, generation after generation of children are taught the folk belief that Vietnamese people belong to *"Hồng Bàng"* family line (a gigantic species of water-bird) and are descendants of the "*Tiên Rồng*" breed ("Fairy" as veneration of an egg-laying species of bird, and "Dragon" a mystical creature from snake and crocodile) (Tran, 1999). It was Mother *Âu Cơ* who laid a sack of 100 eggs ("*bọc trăm trứng*") that hatched into 100 children spreading into different regions of the country. This mythology of the origin of the Vietnamese race is often proudly told in the introduction of the country by Vietnamese Embassies around the world. As a worthy note to take away from these folk stories, the animalistic and totemic nature of the earliest form of Vietnamese religious practice demonstrates the people's belief in human-beings as a part of nature. Little distinction was made between the human and the animal — and by extension, the vegetation, the surrounding environment, the universe.

Vietnamese people also have a set of local worship traditions, worshiping in different places like inside the house (mainly for ancestors), in shrines, temples or palaces. The native tradition of ancestor worship, with altars placed in the most ceremonial place in each household, depicts the respect for origin and sense of continuity. In addition, the heavy dependence of the Vietnamese living on water-rice agriculture from prehistoric time also ignited the belief of respecting and worshiping nature. There still exists a habit especially among older generations in Vietnam that when we wish for something to happen, the opening phrase will be "*lạy trời*", "*cầu trời*" (worship the heavenly god, request the heavenly god's favour). All elements of nature, such as cloud, rain, thunder, lightning, etc. were endowed with magical properties, to become nature deities in charge of forces of nature. This respect of Vietnamese folk beliefs has some commonality with Japanese native religion of Shintoism, where each and every aspect of nature is god (*kami* 神 in Japanese) (Hartz, 2009). There are also spirits living in mountains, forests, rivers and the sea, ranging from those species in river valleys such as water-birds, deer, frogs, snakes, crocodiles (which are gentle in Vietnamese folklore), to plants such as the rice plant, banyan-tree, mulberry tree (Vietnam Embassy, 2017). I still recall the neighbour kids and I for some reason being rather scared of those little shines hiding in secluded places and full of air of superstition. We heard the stories around them and the warning of not to damage or show disrespect to them, "especially banyan-tree" as my mum's note.

Perhaps the most visible difference between Vietnamese folk beliefs and Japanese Shintoism or worshiping traditions in China was the dominance of the *female* figure, from mystical to supernatural to historic characters. Earliest "gods" in Vietnam had no maternal virtues or human characteristics and later on, the worship of Mother *Âu Cơ* of the Viet nation, Mother Goddesses of Heaven, of Forest, and of Water, and numerous heroic figures from different ethnic groups of Vietnam (Vu, 2006). "*Đạo Mẫu*" (Mother Goddess, worshiping female Creators) is in fact classified as a distinct form a Vietnamese folk religion (Ngo, 2010). Beliefs in fertility and worships of female figures reflect the existence of matriarchy in early Vietnam, as the agricultural society possessed "softer" values, stressed harmony, emotions and relationships (Tran, 1999). These feminist values seem to be consistent with the people's wish to live in harmony with Nature, and respect of spirits existing in all living entities as mentioned above. However, later on they were taken over by male dominant values from China, especially after the *Lê* dynasty in the 15th century chose Confucianism as the national religion (Tran *et al.*, 1998).

5B.3.2 *Buddhism*

Compared to Vietnamese folk beliefs, Buddhism is an orthodox religion that was the earliest to be brought into Vietnam from outside. It is among the largest religions of the world and very familiar to Vietnamese people, to the extent that Vietnam sometimes is described as a primarily Buddhist country (General Statistics Office, 2017). This religion was originated from India in the 6th century BC, consisting of the teaching of the Buddha, also known as Gautama Siddhartha. According to Tai Thu Nguyen (2008), historical evidence suggests that Buddhism was first naturally brought into Vietnam through Indian traders and missionaries in the 1st or 2nd century AD (Nguyen, 2008). Then during the late Chinese domination stage, Buddhist influence from the north came to replace the direct southern route from India. The so-called Mahayana (Great Vehicle) branch of Buddhism rapidly spread its influence in Vietnam, at times recognised as the state religion (Nguyen, 2008). Over ups and downs throughout history, Buddhism has remained the prevailing religion in the country and major cultural force impacting people's life. Even for someone who does not claim to follow this religion like myself, I was still taught principles that were stemmed from Buddhism, such as understanding the law of cause and effect to avoid doing evils to others. Buddha also merged into Vietnamese folk stories, such as the tale of the two sisters *Tấm Cám* to help good-natured people. Vietnamese people may not take part in Buddhist rituals at the pagoda, but still go to temples to wish for good luck and fortune. The extension of the opening phrase of "*lạy trời*" (worship of the heavenly god) to add "*lạy Phật*" (worship of Buddha) presents the mix of this orthodox religion with native folk beliefs.

There are several philosophical premises of Buddhism that are relevant in this discussion. First, the Mahayana Buddhism movement advocates not only that everyone is equal, but also human beings and other species are equal. This is different from Western philosophy, where it states that human beings are the centre of the universe, the master of all species (Tran, 1999). All entities have a soul, and thus human beings need to reduce harming and avoid killing. Moreover, an individual's fate in this life is determined by what he or she has done in his previous life and in turn good deeds in this life will be rewarded in the next — this is the law of cause and effect. Another aspect of the teachings of Buddha, which could be seen as negative, is that man was born into this world to suffer. The cause of suffering is craving for wealth, fame, and power, which leads to frustration and disappointment. If a person is attached to a desire, eventually he/she will be controlled by that desire. In order to be free from suffering, man must suppress desires, detach from worldly possessions, enter a state of freedom that is ideal in Buddhist thought, known as Nirvana (Nguyen, 2008).

5B.3.3 *Taoism*

Taoism (or Daoism) in Westerners' mind is very clearly the Eastern philosophy that teaches living simply and living in harmony with nature. Originated from ancient China, this philosophy was based on the writings of *Lao-Tzu*: The Classic of the Way and its Power, or "*Tao Te Ching*" (Chan, 1999). Taoism arose at about the same time with Confucianism (the teaching of *Kung-Tzu*, or Confucius, another ancient Chinese philosopher) but according to the renowned German philosophy Hegel, it is *Lao-Tzu* not *Kung-Tzu* who is the representative of ancient

Eastern ideology. Taoism entered Vietnam as when Confucianism still could not manage to penetrate into the country. The reason was because of its sympathy and similarity with the ancient beliefs of Southern agricultural culture — *Lao-Tzu* (Lão Tử) in Vietnam was known with the image as a Southern farmer riding a buffalo, unlike the formal portrait of *Kung-Tzu* as a Northern (i.e. Chinese) mandarin (Tran, 1999). It may not be realised by many Vietnamese people but images of the Gods of Taoism are used in temples and pagodas throughout the country. Most home worship the "Kitchen Gods", the three Taoist deities that protect the household. Many of Vietnamese festivals, including *"Tết"* (celebration of our lunar new year), have a Taoist tradition (Vietnamese Culture, 2018).

According to *Lao-Tzu*, human race is a microsome of the universe, representing the link between time (present and past) and space (sky and earth), and providing the balance between them (Cooper, 2010). There is an existential organising principle of the universe (so called Tao 道 in Chinese, or *"Đạo"* in Vietnamese, meaning the "Way"), forming an interconnected whole between everything in nature. In this universal cosmic structure, natural laws, which are also known to and practised by Vietnamese people, all apply. I personally have poor recollection of learning about *Lao-Tzu* or Taoism at school in Vietnam, but I practise these natural laws often: The Five Elements theory (gold, wood, water, fire, soil), *"Phong Thủy"* therapy (*"Feng Shui"* in Chinese arrangement to maximise positive life energy), the *"Âm dương"* law (*Yin-Yang* — two opposite yet complementary energies). A person can be with the "Way" by living in harmony with nature and all its transformations, and by changing, adopting and assimilating to these, in order to gain eternal life (Chan, 1999).

Another tenet of Taoism is that the universe already works harmoniously according to its own way and therefore human beings are not to interfere with the spontaneity of the "Way" or alter it by any means. An example of how the "Way" already works well is the harmonious change of the four seasons year after year. If a person exerts his will upon the world, he or she will disrupt the existing cosmic structure; it is going against the way of life. This philosophy is expressed by *"Vô Vi"* (*Wuwei* in Chinese) meaning no action, non-interference. This is not about laziness or complacence, but regarding letting things take their natural course in accordance with the "Way", ultimately no forcing or taking any action opposing nature (Cooper, 2010).

In addition, Taoism also believes that if one looks at life and thinks about things in the right way, one will be happy. By urging people to "manifest the simple, embrace the primitive, reduce selfishness, have few desires" (Ammi, 2013), *Lao-Tzu* demonstrated one of the "three treasures" of Taoism: moderation/ simplicity/ frugality, as noted in his work *Tao Te Ching* (verse 67).

The above are among the major premises of the Taoist philosophy that can impact on a person's attitude towards nature and life. As noted earlier on, Vietnamese people do use many images, concepts, and other elements of Daoism in daily life, worships and traditional celebrations, yet the pure form of this philosophy may not be easily recognised by the common people; instead elements of it have been absorbed into the Vietnamese folk religion (Turner and Salemink, 2014, p. 240).

5B.4 A unified system of values and beliefs to drive "sustainability"?

It has been long established that religious and philosophical traditions are deeper underlying mechanism that impact on attitudinal and behavioural responses (Bond, 1996). This piece of work started with the objective to provide an account of Vietnamese traditions and evaluate whether these values are in fact the internal capability among Vietnamese people to pursue a sustainable way of life. So far three traditions have been reviewed, consisting of Vietnamese indigenous beliefs and the foreign-originated Buddhism and Taoism. It is noted that there is great tolerance among the Vietnamese towards different religions and beliefs (Lewis *et al.*, 2009); those native and adopted values have all been coexisting peacefully in the society here for centuries. They are interconnected and have mutually influenced each other through a wide variety of rituals and in overlapping cosmologies and pantheons (Salemink, 2008). Buddhism when brought into Vietnam became more feminine and included female venerated figures under the influence of Vietnamese indigenous culture (Tran *et al.*,

1998). Meanwhile, many aspects of Taoism merged into the native Mother Goddess religion and some other local religious sects (Turner and Salemink 2014). Vietnamese culture is after all a syncretic amalgamation of a variety of values and beliefs from different origins to indeed form a *unified* system. The overall evaluation of Vietnamese traditions as the drive for a sustainable lifestyle therefore should have a holistic approach, as aggregated below.

(1) The long-established occupation of water-rice agriculture with the need for settlement by Vietnamese people meant long-term dependence on nature. Vietnamese agriculturists wished to have a good relationship with nature, understanding and respectful.

(2) Vietnam as an agricultural civilisation possessed "soft", feminine values, focused on affection and emotion, with people favouring the weaker and loving a good cause. It is not just a relationship with nature but a loving relationship with harmony. Vietnam scores as a feminine society on the Feminism (vs. Masculinity) dimension in Hofstede's model (Hofstede Insights, 2018), *whilst* feminine values have been known to be associated with environmentalism (de Groot and Steg, 2007).

(3) The folk belief that human beings are an integral part of nature, the Taoist folk belief in no separation to be made between human and other entities of nature, "non-killing" nature of Buddhism because all species are equal, plus all entities having a soul or spirit in both folk beliefs and Buddhism — all lead to our appreciation of co-existence with others in nature, for mutual benefit.

(4) The totality and holistic view of Taoism states that all elements exist as an integral whole and not in separation. Human beings are not to go against the course of nature, breaking this existential order and structure. For example, the harmony with nature is disordered as the result of man-made global warming. A naturalistic way of life, making no impact, no footprint is advocated in this philosophy.

(5) The focus on restraining desires in both Buddhism and Taoism is interesting as the purpose ranges from avoid suffering, get a feeling of freedom to happiness. Happiness is not how much you own but decrease what you want in Buddhist way. Both traditions advocate

man to want less and not rely on materialist possessions, to live "simple". In this respect, the wisdom is close to the concept of "Voluntary Simplifier" by some scholars in the West (McDonald *et al.,* 2006). Vietnamese is characterised as "restrained" on the restraint versus indulgence dimension in Hofstede's framework (Hofstede Insights, 2018). Restraint of needs and want is the sustainability idea of well-being within limits!

(6) Vietnamese have a long-term view as the result of settlement in agricultural civilisation. The ancestor worship tradition in folk beliefs expresses the sense of continuity. The long term view means it is not just for own but for future generations. Buddhism also believes in leave good virtues for the next life, next generation. This mentality advocates concerns for future generation, i.e. a sustainability issue.

(7) Group mentality (or collectivism) is the natural outcome of settlement in groups in water -rice agriculture. Twinned with limited resources in difficult time, sharing is a traditional way of life here, which has recently been heavily promoted in contemporary pro-environmental literature (Collaborative Economy, 2017). Moreover, history of hardship and limited resources also means Vietnamese have no issue with reuse, repair, fix, lengthening the life of a product. This is the idea of another sustainable way of life — the Circular economy (Pearce and Turner, 1989). I used to question whether it was pure thrifty when my grandma never stopped her rigid effort to collect all little plastic bags lying around, try to find most use for them, then reuse again and again. She passed away decades ago but after all, maybe she was the one who could foresee a future way of life, not me!

5B.5 Concluding remarks

It is by no means a sustainable Earth that we are living in today. The problem of damaged and depleted natural environment can be more serious in some countries than others. Vietnam as a country was left in devastating condition after the war and it is still going through a period where "sacrifices" of the natural resources and environment sometimes are made in the name of economic development and growth. As a positive note, Vietnamese people already hold a system of religious and philosophical traditions that advocate ways of thinking, ways of living towards the sustainable target. These traditions should also be long standing rather than eroding in the course of modernisation because they are strengthened by the collectivistic nature of the Vietnamese society, through the work of social agents such as family, school and the government. The next step for the country involves how to turn into action, close the attitude-behaviour gap. It must be noted that the national State can play the role of another "pillar" of institution, in addition to that of culture and traditions, to help the nation and people to channel to the desirable behaviour. The State is to provide security and legal frameworks, reduce superstition that could stem from some religious beliefs so as to promote a hope appeal in this good cause rather than fear appeal. After all, this is only one example of a nation with traditions and values that link with pro-environmental attitudes. Other countries in the region and probably different parts around the world may share some commonality in their value system or have the wish to adopt/ return to Eastern values. This is a good sign for a hopeful future of a more sustainable Earth.

Chapter 6

AFRICAN DEVELOPMENT
AND MANAGEMENT

6A — Development in Nairobi: Three
Into One Does Not GO!

Collins S. Makunda

University of Nairobi, Nairobi, Kenya

collins.makunda@gmail.com

6A.1 Introduction

Cultural heritage represents an opportunity for defining a locally appropriate mode of architectural expression. In a globalised world, driven largely by neoliberal doctrine and ideology, a "universal" cultural mode of expression has become pervasive. In turn, diluting the unique cultural identities that are the defining characteristics of the diversity that is at the root of the richness of all the world's varied communities. This is keenly apparent in the context of the housing boom in Nairobi. Over the past decade, low-density suburban districts have experienced rapid vertical densification resulting in a preponderance of high-rise apartment blocks where single-dwelling units historically held sway. The emergent new apartments are "modern" in the universal sense of the word representing the architectural expression of flows of global capital. Little of the buildings' design, form and layout echo the local cultural context. Situated in the global South, they differ very little from architectural expression in the global North and bear little resemblance to local and traditional modes of architectural expression. Yet, an opportunity exists to celebrate local culture and mine it for locally appropriate modes of architectural expression. Indeed the 2016 *Ngorongoro Declaration* in Tanzania (UNESCO, 2016b) was a call to recognising that culture is a crucial ingredient in the path towards achieving sustainable development; and, in this case, particularly on the continent of Africa.

6A.2 A rapidly transforming city

Nairobi is the capital and largest city of Kenya as well as the largest economy in East Africa. It was established by British colonialists more than a century ago, in 1899, as a railway encampment during the construction of the Kenya–Uganda Railway (Halliman and Morgan, 1967, p. 99; Freund, 2007, p. 79). However, the initial plan for the town as a railway depot was prepared by Arthur Church in 1898 (Hill, 1976 cited in Teckla *et al.*, 2016, p. 76). From a settlement that covered little more than seven square miles in 1901, it had grown into a city covering 266 sq mi (689 sq km) by the time of independence in 1963 (Halliman and Morgan, 1967, p. 100). Concurrently, the city's population had mushroomed from 11,512 in 1906 to 342,764 by 1963 (*ibid.*). While the area covered by the city has remained virtually unchanged since independence (269 sq mi/696 sq km according to the Kenya National Bureau of Statistics, [KNBS, 2015a]), Nairobi's population has increased more than ten-fold over the 50-year period since. As of the 2009 population and housing census, the city's population was 3.138 million (KNBS, 2015b). Its current population is now estimated at four million (World Bank [WB], 2014) accounting for approximately 9% of Kenya's total population.

The city of Nairobi, as the primate city of the country, is also the epicenter of the rapid transformation that Kenya has been undergoing since independence. It is the locus for the rapid population and economic growth experienced by the country, especially in the new millennium. The country's economic growth rate (GDP) has been consistent at 5% or higher since the third quarter of 2014 (KNBS, 2017). And in tandem with other cities and towns in the country, Nairobi accounts for the almost 30% state of urbanisation of the country (World Bank, 2016). The level of Kenya's urbanisation is expected to increase to 44% by 2050 (Worldometers, 2017). Along with the rapid urbanisation, the economic transformation of the country is exemplified by the rise of a middle class estimated to have increased countrywide from 3 million to 5.5 million, and trending upwards, between 2006 and 2012 (Shah and Ruparel, 2016). Most of this emergent class is domiciled in urban areas and particularly Nairobi.

The growth of the middle class has been linked to the increase in construction activity in the city, which has witnessed a construction boom in the real estate sector over the past decade (Hass Consult, 2017). Mega malls have been developed and so have high-end apartments in prime residential areas. In addition, a new consumption culture closely allied to globalisation has emerged in the city. Car ownership has increased and a number of international brand name establishments such as *Carrefour*, *Subway*, *KFC*, *Dominos Pizza*, and *Uber* have recently opened their doors in the city presumably to capture the increasing disposable income that a growing middle class represents.

However, the transformation of the city has arguably been most keenly felt in the real estate sector particularly in the rapid development of high-rise apartment projects. This raises many questions of sustainable development especially regarding the processes and outcomes of current trends.

6A.3 Towards sustainable development

The Brundtland Report, *Our Common Future*, considered the economic, the social, and the environmental as integral components of sustainable development (World Commission on Environment and Development [WCED], 1987) while the United Nations Educational, Scientific and Cultural Organization (UNESCO) considers the cultural dimension as equally important (UNESCO, 2016a). In 2016, the *Ngorongoro* declaration made it clear, especially for the continent of Africa, that cultural heritage had to be an integral component in any attempt at attaining sustainable development (UNESCO, 2016b).

What is easily lost in discussions on sustainable development is the importance of the cultural dimension.

6A.4 Unsustainable development

Three decades since the publishing of the Brundtland Report, the economic dimension espoused therein has arguably overshadowed all the other components integral to sustainable development. This is evident in the housing boom that the city of Nairobi continues to experience. Taking the example of the western suburbs of the city; single family bungalows on half to three quarter of an acre plots, constructed towards the end of the colonial era in the 1950s and 1960s, are being demolished and replaced by townhouses, mid-rise housing, and high-rise apartments of up to 13 floors. This intensification of land use from low density residential to high density residential has not been accompanied by requisite infrastructure. The sewer system, drainage, water reticulation and road network in these suburbs was designed for low density housing. However, these systems have not been upgraded despite more than a decade of high-rise apartment development.

6A.5 The case of Kileleshwa neighbourhood

Kileleshwa, one of the neighbourhoods in the western suburbs of Nairobi, exemplifies the magnitude of the transformation. It is a neighbourhood that has historically been referred to as a leafy green suburb of Nairobi planned according to garden city principles (Freund, 2007, p. 79). This is because, for at least half a century, it has been characterised by low-density single family housing on individual three quarter of an acre plots with plenty of trees and greenery both on individual compounds and shared public streets. But in the new millennium, as the urbanisation pressure continues unabated and the middle-class demand for housing close to the Central Business District (CBD) rises, the single-family units, mostly bungalows, are rapidly being replaced by high-rise apartment blocks (Makunda, 2017a, 2017b). As a testament to the power of the economic dimension in this context, the apartments being developed far exceed, by more than triple, the stipulated height limit of the residential area. *Kileleshwa* neighbourhood is zoned as a residential area with a height limit of no more than four floors (Architects Forum Kenya, 2015). The foregoing notwithstanding, currently, where there used to be trees and their canopy as the dominant feature of the residential area, a haphazard skyline of apartments now dominates the vertical space along with perpetual traffic at ground level. The apartments have also increased the hard surface plot coverage due to the larger plinth area of their foot prints, far in excess of the regulatory plot ratio. And, an immediate consequence of this, apart from the trees that have been felled to make more room for the larger built structure, has been the increase in surface runoff whenever it rains leading to floods in the lower area of the neighbourhood, with water deep enough to submerge a passenger vehicle. The flooding is further exacerbated by a drainage system not built with the capacity to handle the volume of water generated by the expanding concrete jungle (Makunda and Edeholt, 2016).

6A.6 Cultural architecture

It is arguable that in the juxtaposition of the two terms; "culture" and "architecture" is embedded the idea of architecture encompassing culture. Thus, for example, modern architecture could be seen as architecture that bears the aesthetic sensibilities of modernism both in terms of form and disposition with its allegiance to efficiency and minimalism and use of materials made possible by the industrial revolution.

In the context of Nairobi, scholars have discussed cultural architecture in relation to historic buildings located in the city's Central Business District (CBD) noting the intermingling of foreign influences and local cultural heritage, including locally derived building materials, in the architectural expression of the built form (Teckla *et al.*, 2016).

However, the architectural form adopted for residential buildings in the city lacks even a modicum of acknowledgement of the rich local cultural heritage. This is keenly apparent in the apartment blocks currently being developed in the residential areas of the city that, arguably, properly fit into an international aesthetic in form and appearance, as discussed further below. This denies the city an opportunity to imprint its local identity on the character of the housing being produced for its residents.

6A.7 The emergent architecture of Kileleshwa neighbourhood

The architecture of the apartments being developed as living solutions for Nairobi's residents lacks even the faintest of echoes of the local cultural heritage and context. This is particularly apparent in the formal design, layout and style adopted for the apartment forms, copies of which could easily be found in any number of cities situated in the western hemisphere (Makunda, 2017b). In fact, the new apartments' exotic characteristics are marketed as a selling point with the terms "western" and "modern" used as signifiers of a mode of living that is promoted as being aspirational for those with the economic means to buy into the advertised lifestyle.

In many ways, the emergent architecture of the *Kileleshwa* residential neighbourhood, in the suburb of Nairobi, as exemplified by the typologies of the newly developed apartments is quintessential of a modernising urban residential area rather than an ecologising one (Latour, 2012/2013). It is an expression of a globalised form of architecture that is leaving its unmodified imprint in varied contexts both in the global North and global South. And, it represents a lost opportunity for a culturally appropriate mode of architectural expression.

The gate keepers of property development in the city are best positioned, through the property development approval process, to insist on locally appropriate cultural architecture rather than a globalised one if they are to ensure the safeguarding of the uniqueness of the local context.

Property developers, left to their own devices will opt for tried and tested architectural forms, layouts and designs that minimise their exposure to risk while maximising their opportunity to profit from the endeavour. For it is all too convenient for them, given a choice, to fall back on using readily available templates of apartment designs. However, what is lost in the process is the opportunity to define a locally relevant form of architectural expression that not only celebrates and safeguards local African cultural heritage but also allows room for indigenous cultural knowledge, honed over millennia, to be deployed in developing creative living solutions that would not only foster local pride and a culturally appropriate mode of living but also contribute to sustainable development in drawing a majority of its resources and inspiration from the local context.

6A.8 An ecological perspective

From a purely economic, market-driven point of view, the apartment developments can be considered a success; first, as signifiers of high financial investment in the real estate sector thus contributing to high economic growth, and, second, as representing an ideal situation where the market need for housing, albeit only for a particular socio-economic class, is being met. But this is a limited lens. From an ecological perspective, where the economic, the social, the environmental and the cultural dimensions are taken to be equally important (World Commission on Environment and Development, 1987; UNESCO, 2016a), the apartments as currently conceived and developed cannot be considered a success because they privilege one dimension at the expense of many other dimensions: The social and the cultural dimensions are neglected in so far as inadequate shared spaces are provided, and cultural practices and expectations are not taken into consideration in the apartment design layouts and finishes. The environmental dimension is disregarded as tree cover is destroyed to make room for apartment blocks; and sustainable strategies in relation to energy, water and waste management are ignored in favour of the maximisation of profits.

6A.9 Conclusion

If sustainable development is to be achieved in the face of rapid population growth and urbanisation in the global South, the market-driven economic dimension as a singular criteria and metric for measuring development, has to be abandoned in favour of a more robust multi-dimensional approach. An ecological approach, that factors in all the key dimensions in equal measure ought to be adopted as an exercise in balancing competing but also necessarily complementary dimensions of sustainable development. It may not be the easier alternative to adopt nor does it lend itself to easily measurable indicators, but it is the necessary and responsible path to follow if sustainable development, understood as, "the use of resources in such a way as to not compromise the ability of future generations to use the same resources," (World Commission on Environment and Development, 1987) is to become an actualised reality, particularly in relation to achieving the Sustainable Development Goal (SDG) 11 on sustainable cities and communities, that expresses the aspiration to: "Make cities inclusive, safe, resilient and sustainable (United Nations [UN], 2016). And as aptly expressed in the 2015 EEUM Declaration, recently ratified in the Montreal Design Declaration during the World Design Summit, the driving mantra of creative actions ought to be: "Creating a world that is environmentally sustainable, economically viable, socially equitable, and culturally diverse." (World Design Summit, 2017).

6B — A Tale of Two Theoretical Cultures: Expanding the Application of Habermas' Communicative Action and Post-colonial Theories in Management Studies and Practices in Africa

Sharif M. Khalid[*,‡] and Adeyinka A. Adewale[†,§]

*University of Sheffield, Sheffield S10 2TN, UK
† University of Reading, Reading, UK
‡s.m.khalid@sheffield.ac.uk
§adeyinka.adewale@henley.ac.uk

6B.1 Introduction

A plethora of post-colonial or Commonwealth literature exists within the academic sphere. Although not definitive within a specific continental space, it very much spans a wide scope across all continents of the world, race, ethnicity religions, class groups and even with the cosmopolitan empires. For its ambivalent nature, the colonialised assumes some status of freedom yet colonised. Edward W. Said, Homi K. Bhabha, Frantz Fanon, Robert J. C. Young, Gayatri Chakravorty Spivak among others are leading scholars that have dedicated their academic toils to post-colonial studies. However, for the purpose of the paper, much is going to be anchored of Said, Fanon and Bhabha where a juxtaposition with Habermas's Theory of Communicative Action should offer a by-product of a hybrid of Post-colonial-Habermasian theory. This should lead to an applicable theoretical framework within the management, business, academic and practitioner field in Africa.

The yawning business and theoretical model for the emerging or developing world makes business and education erratic, nondefinitive and lethargic; above all experimental and destitute. As much as the BRIC countries of Brazil, Russia, India and China have been touted as model states within the colonised world, an uphill debate in pointing to a functional model in these nations still exists. China is often given a special treat as to having a defined model but even then, there are many who are of the assertion that China is borne out of globalisation and not built on the so called "Chinese model".

Victorian Britain stands as a classic case of the model of cultural and economic renaissance for a thriving society. The Thatcherite and Reaganesque models have left indelible prints of economic concepts for the Conservatives in Britain, the Republicans in the United States, including the wider political domains in these countries. However, a stark difference manifests itself in Africa. Though seeking much inspiration from communism and exiting with it, Nkrumah triggered a model in post-colonial Ghana. Even recent independent nations like South African often referred to as a somewhat economic mecca of Africa still sit on what cannot be described as an African model.

Both the twilight of the 20th century and the genesis of the 21st heralded a period of African global renaissance within the sonic modern space. This ranged from fashion to cuisines. But conspicuously missing in the equation is a business, and to a large extent a management model. This however could be attributed to the nonexistence of a robust theory with African ethos embedded within it. However, from Abuja to Juba, efforts are being made. A manifestation of this is what is being done in institutions like the Tony Elumelu Foundation in Nigeria whose aim, through its Africapitalism Institute, is to promote what they term as "Africapitalism". By Africapitalism, they refer to "an economic philosophy" where the African private sector has the power to transform the continent through long-term investments in strategic sectors, creating both economic prosperity and social wealth."

(Elumelu, 2018). This clearly does not engrain in itself any African value or practices. Also evident, is the institutionalisation of mad-in-Ghana/African clothes on designated days — usually Fridays — in African cities like Accra. Sadly, so, this noble concept defeats itself, as most of the so-called African/Ghanaian fabrics are not made in Ghana as otherwise wished. All of these examples as enumerated points to the unquenchable thirst for an "Africanisation" of modern Africa. It is indeed reflective of the marginalisation of Africa in management and leadership discourses (Nkomo, 2011). The recognition of this marginalisation should be one seen in both the colonial past and present (Ahluwalia, 2001). However, according to Frenkel (2008), multinational corporations serve as a positive conduit for the importation of best practices into Africa. A sort of liminal connotation that stifle potent growth of the colonised.

The impetus to this paper is therefore hinged on what Post-colonial theory actually is and how it could be modelled among other theories furnishing furnish it with a rather fit for purpose applicability within the African Management space, with African cultures as key variables. This is done with due cognizance to the fact that the post-colonial theory is a clone of post-modernism and post-structuralism, both of which are non-African in nature. To do this, the post-colonial theory shall be analysed analytically, drawing inference from Said, Fanon and Bhabha. This shall then be juxtaposed with Habermas' Theory of Communicative Action to permit the application of a hybrid theory for management studies and practices in Africa or otherwise.

Edward Said was a Professor of Comparative Literature at the Columbia University in the United States. Most of his works focus on Western hegemonic presence within subaltern cultures. His seminal piece, Orientalism (Said, 1987) is applied to this study.

Frantz Fanon was an Afro-Caribbean psychiatrist and philosopher trained in France and practiced in Algeria. After Algeria's independence from the French, he served shortly as Algeria's ambassador to Ghana. He is best known for his work in psychological effects of colonialism. Black skin, white mask (Fanon, 2008) and The wretched of the Earth (Fanon and Farrington, 1969) stand as his greatest contribution to post-colonial studies. Homi K. Bhabha, somewhat inspired by Said expands on Said's post-colonial works. His magnum opus The Location of Culture

(Bhabha, 2004) offers vivid interpretation of colonial legacies and their implications on the colonised. Jürgen Habermas on the other hand is a 21st century German philosopher. He belongs to the Frankfurt school of thought known as critical studies. His theory of communicative action has been and continuous to be applied in areas as media, journalism, accounting and management studies among others.

6B.2 The post-colonial theory

According to Bhabha (2004, p. 9) "Post-coloniality, for its part, is a salu-tary reminder of the persistent "neo-colonial" relations within the "new" world order and the multinational division of labour" (Emphasis added). There exists a plethora of definitions to post-colonialism, but the common thread that runs across all definitions is marginalisation and indoctrination — what Bhabha describes as mimicry, liminality and hybridity leading to ambivalence. Post-colonialism, as referred to in this research, is one devoid of emotive tendencies but from an approach aimed at distilling an application under its current nuanced form. There exists no post-colonial theory that has no reference to Western writers, thinkers and theory. Since its inception this is a loosely discursive theory that lends a lot of credence to Western theories like post-modernism, post-structuralism and the like. Foucault, Gramci, Derrida, Shakespeare among others have been borrowed into post-colonial theory. All of these make it difficult to be actually "Africanised". Said in his seminal piece on post-colonialism through the illustration of the "Orient" and the "Occident" — terminolo-gies Henry Kissinger in today's terms would describe as the "developed" the "developing" sets an intellectual pace for post-colonial studies.

Although Said's reference to the Occidental world is in direct reference to Asia and the Middle East, it is, however, much representational of the colonised and now post-colonial world. According to Said (1987, p. 40):

> "[t]here is no way of putting this euphemistically. True, the relationship of the strong to the weak be disguised or mitigated, [...]. But the essen-tial relationship, on political, cultural, and even religious grounds, was seen-in the West, which is what concerns us here-to be one between a strong and a weak partner".

Stemming from this, Homi K. Bhabha, posited that the sort of existential relationship that existed between the developed and the developing world in otherwise the Orient and the Occident did produce some by-products viz., mimicry, hybridity, liminality and ambivalence. The entire concept of *hybridity* is amply expressed in simple terms as:

> "They are also signs of a discontinuous history, an estrangement of [a people]. They mark the disturbance of its authoritative representation by the uncanny forces race, sexuality, violence, cultural and even climatic differences which emerge in the colonial discourses" (Said, 1987, p. 161).

Through this process of hybridity, a substantial amount of originality is lost. By hybridity, Bhabha refers to a mélange of Western and African or third-world doctrines, where Western doctrines appear dominant over African doctrines. A result which leaves Africa in a complex state of not belonging to either doctrines in full but laying in-between. This was made possible through colonisation and now globalisation. However, it is Fascinating to note that, interest is the prevailing factor here. A classic example of a case of hybridity is the practice of Indirect Rule and the Policy of Assimilation by the British and the French respectively during the colonial heydays. Through the policy of Assimilation, colonised Africans were converted into French Africans whereas Indirect Rule was a colonial system of British rule where Africans were ruled through their local chiefs. What these practices left was a system that was neither British nor French and definitely not African. Hybrid products such as these have left Africa still grappling with what have been christened "the African Self", a popular phrase that heralded Africa's independence in the 1960s.

> "Mimicry [on the other hand] reveals something in so far as it is distinct from what might be called an itself that is behind. The effect of mimicry is camouflage [...] It is not a question of harmonizing with the background, but against a mottled background, of becoming mottled-exactly like the technique of camouflage practiced in human warfare." (Lacan, 1977, Cited in Bhabha, 2004, p. 121).

Mimicry therefore makes these hybrid products of colonisation appear to represent what they are not thereby creating a lost representation of the "African Self", resulting in ambivalence. With the hegemonic presence of the coloniser within the colonised space, it leads to an inescapable mental space for the colonised. This, Bhabha refers to as liminality. Fanon however looked at the hegemonic presence of the West within the colonised space as one that has left a psychological scar. A scar the places the African as a second-class citizen to the rest of the world.

6B.3 Habermas' communicative action theory

Jurgen Habermas, a philosopher of the Frankfurt School is the originator of the Theory of Communicative Action. Within the Theory of Communicative Action are concepts like an "ideal speech situation", "lifeworld", "steering institutions" and "steering mechanism" (Habermas, 1984, 1990; Laughlin, 1987).

From Habermas' perspective, an ideal speech situation interlocutor within a communication process must each have the same level of advantages with no cohesion on the part of any. Meaning they all must exist on the same level playing field. If this is rationalised into both today's terms of colonisation and that of previous years any form of cohesion must be frowned upon. This touts equity, equality and the sort of justice ideal democracy has since the last 200 years or so, been struggling to achieve. Most of Habermas' position of an ideal speech situation comes across as an abstraction and would be considered idealistic. An attempt at its practice could archive some worthwhile miles between the two worlds of the developed and developing. Steering intuitions connotes the governing institutions of society i.e. courts, security and public sector establishments. These steering institution produces governing rule or codes for society referred to in Habermasian terms as the lifeworld or society. These codes or rules produced by the steering institutions are what is known as the steering mechanism, i.e. national constitutions, laws, management codes etc. Interestingly though, all of these codes and instructional establishments are but colonial legacies, artefacts and a sole importation of Western practices with little or no recourse to African beliefs and traditions into an African system. All of which are conveyed through the globalisation thread. Thus, Africa seems to have accepted its own appreciation of the 21st century colonisation masquerading as globalisation. Should there exist an African concept within the global mix?

6B.4 The interconnectedness between Habermas' theory of communicative action and the post-colonial theory

Habermas' theoretical representation and that of most if not all post-colonialist have certain commonalities. They at least are on firm grounds towards minority societies, marginalised groups; political and economic structures; and desirous towards a certain SOCIAL/COMMON GOOD as well as modernity, all in the face of equity and equality.

> "In attempting to elucidate the concept of rationality through appeal to the use of the expression 'rational', we had to rely on a *preunderstanding* anchored in modern orientations. Hitherto we have naively presupposed that, in this modern understanding of the world, structures of consciousness are expressions that belong to a rationalized lifeworld and make possible in principle a rational conduct of life. We are implicitly connecting a claim to a *universality with our Occidental understanding of the world*" (Habermas, 1984, p. 44).

Bhabha (2004, p. 246) amplifies the connection between Habermas and the post-colonial theory as:

> "To bend Jürgen Habermas to our purpose, we could argue that the post-colonial project, at the most general theoretical level, seeks to explore those social pathologied- 'loss of meaning, conditions of anomie'-that no longer simply 'cluster around class antagonism, [but] break up into widely scattered historical contingencies'".

Through the theory of Communicative Action, Habermas is explicit on establishing an ideal speech situation aimed at communicative satisfaction between two or more interlocutors. In this case between the developed and the developing world.

Also, both Habermas and the post-colonial theory draw on discourse analysis for case making and theory formation. Adding to the presence of Habermas in post-colonial discourses, Quayson (2000, p. 34) writes:

"Mbembe brings an innovative note to the study post-colonial African politics in his combination of [other critical theorist] and Habermas." The two dimensions of the Black man viz., one with his fellow blacks and the other with the whites, Fanon (2008), creates a somewhat constant disadvantage in the continual post-colonial existential relationship between blacks and whites-in otherwise the Privileged and the Underprivileged Worlds. As Fanon (2008) believes in two poles of the black and white world within a psychological frame, so does (Habermas, 1984).

6B.5   The new post-colonial theory

As discussed above, the post-colonial theory is pregnant of many theories and concepts making it eclectic in nature with little to do with Africa expect for a wider marginalised group of people among which includes Africans. It is therefore evident that the post-colonial theory in its very nature stands as inadequate for an exclusive and critical examination of a true African management concept. Expect for a discursive interrogation of phenomena, it comes as no "jewel on the crown" in its entirety without furnishing support from other theories, begging the quest for a more elaborate ethos engrained with African doctrines. To do this, some noble yet practical African practices such as Ubuntu and Noboa should be allowed prominent and active space within post-colonial theory. Ubuntu is a South African concept of a "we feeling", sharing, care and togetherness, whereas Noboa is Ghanaian traditional practice of mutualism. This is most prominent among local family communities, through which families serve as farm hands, taking turns on each other's farms throughout the entire farming season in mutual formats. The interesting nature of these concepts among other traditional African practices is that: they are egalitarian in nature and permeate almost all of Africa, if not all. Translating these concepts into business practices would create some striving grounds for the germination of African business models and case studies for academic and practitioner purposes. Moreover, a fertile ground would be established for these concepts to become embedded within the post-colonial theory, thereby Africanising it. Nevertheless, all of these should be done with due cognizance to the fact the post-colonial theory is not sacrosanct to Africa but that of the globally colonised and marginalised. This means that it may have to attract either a prefix or suffix to grant it an African exclusive. Therefore, it might as well maintain its current state yet aspire pragmatism. Its pragmatic nature should be seen as one that will permit reflexivity and flexibility in usage, which of course it seems. The reflexivity and flexibility of the post-colonial theory should be one that would witness the introduction of traditional values by any of the geographies belong to the post-colonial family to suit its exigent moments. Say for instance Africans introducing Ubuntu, Noboa among other African doctrines into the post-colonial theory, making it exclusive for Africa and the purpose of Africa.

6B.6 Applying Habermas and the post-colonial theory in Ghana's mining industry: The stakeholder accountability and transparency dimension

Ghana's mining industry attracted foreign interest as far back as 1471 (Dashwood and Puplampu, 2010). Owing to the high presence of gold minerals, the country then was referred to as the Gold Coast. As of the time of writing, the official website of the Ghana Chamber of Mines, a mining industrial body pegged the number of registered represented mining companies for its level A category at eight (8), category B at one (1) and exploration companies at seven (7). By represented members, the Chamber refers to mining companies in commercial production whereas exploration companies are representative of those prospecting and exploring for gold mineral resources. As represented on the website of the Ghana Chamber of Mines, over 90% of registered companies with the Chamber are foreign owned and not indigenous Ghanaian companies. Despite the massive presence of large-scale mining operations in Ghana, there is a low level of employment rates for indigenous Ghanaians within the sector (Garvin *et al.*, 2009). Coupled with this problem is the recent presence of illegal foreign miners, mostly of Chinese origin. Both large-scale and small-scale foreign and local mining activities have led to environment degradation, pollution, capital flight, human rights issues, and above all poor accountability and transparency to stakeholders. Stemming from this, we seek to apply a model of corporate accountability and transparency of the Ghanaian mining sector through the lenses of Habermas' communicative action

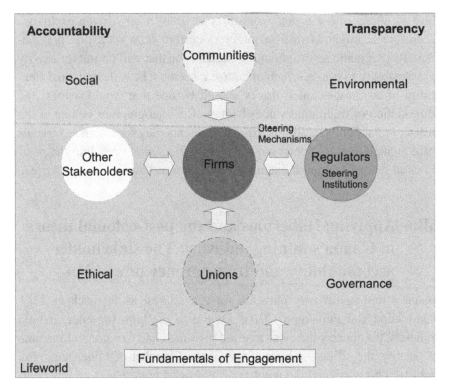

Figure 6B.1.   A Habermasian Model of Corporate Accountability and Transparency for The Ghanaian Mining Industry

theory and post-colonial theory. Here, cognizance is given to the existential relationship between the Ghanaian mining industry and that of multinational mining firms with operations in Ghana. This is done in order to create the possibility of applying similar synergies and commonalities between the two theories in question and their possible application within an African Industrial setting.

Figure 6B.1 depicts an application of Habermas' Theory of Communicative Action and the post-colonial, aimed at exuding corporate accountability and transparency within the Ghanaian mining industry.

Figure 6B.1 depicts a utopian model of corporate accountability and transparency of the Ghanaian mining industry. The entire square containing all the elements within the diagram depicts what Habermas termed as the Lifeworld. He refers to the Lifeworld as a sort of ideal world for the good of all of mankind. The horizontal box labelled fundamentals of engagements with three arrows pointing up is representative of Habermas' conditions of an "ideal speech" situation viz.,

1. All must work towards a common goal.
2. The process should be devoid of coercion.
3. All should be on the same level playing field.

The circle in the middle labelled "firms" is representative of multinational mining firms operating in Ghana. The rest of the four circles surrounding the firms depict Communities, Mine Worker Unions, Regulators and Other Stakeholders of the mining industry. The double-headed arrows pointing from both sides of each circle showcases the interflow of practical actives and interconnectedness of stakeholders within the mining sector. An ideal interaction of these stakeholders and multinational firms should therefore result in a sort of accountability and transparency tangibles through social, environmental, and ethical and governance practices of the sector.

Figure 6B.2 shows corporate accountability and transparency of the Ghanaian mining industry by exuding the effects of the existential relationship between the Orient and the Occidental world (á la Said). The two circles depict the Occidental and Oriental worlds, in otherwise the privileged and underprivileged worlds with the big circle representing the Orient whereas the smaller one represents the Occident. The two arrows pointing down to the circle anchors both the colonial and post-colonial periods. The intersection between the two arrows contains elements of the relationship between the Orient and the Occident namely: Commerce, Slavery, Colonialism, Imperialism, Neocolonialism and Internationalisation. The circle labelled Orient contains the resultant elements of the existential relationship between the Orient and the Occident. Bhabha labelled these elements as: Hybridity, Liminality, Mimicry and Ambivalence. The vertical box represents the multinational world of mining. Accountability and transparency is therefore gauged through the commercial interaction between these two worlds of the Orient and Occident in a form of social, environmental and ethical reporting and practices of the industry and how this is transmitted to the theming stakeholders of the industry in a form of accountability and transparency.

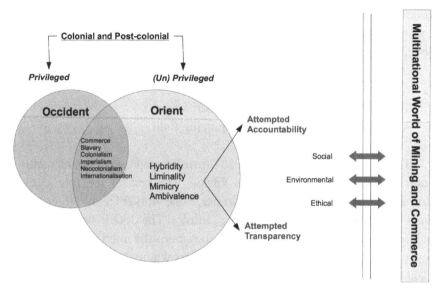

Figure 6B.2.   A Model of Corporate Accountability and Transparency Through the Post-colonial Theoretical Lens

6B.7 Conclusion

From the above, it is clear that a construction of an independent post-colonial theory within an African lens, one imbued with African traditions and doctrines would be a herculean task. This is so because the very foundation of a post-colonial theory is built on Western ideals with inspirations from Western utopian philosophers. Therefore, to attempt a reconstruction of the post-colonial theory will amount to a deconstruction of the current theory leaving behind the foreign concepts within it. This therefore leaves all post-colonial writers and theorists at a cross-road — one that would either create way for the deconstruction and reconstruction of the *post-colonial* theory within an African or post-colonial frame; embracing concepts like "Ubuntu" (we feeling) and "Noboa" (mutual assistance). "Ubuntu", a popular Southern African tradition of togetherness and "Noboa", a Ghanaian traditional farming practice of mutual assistance are potent traditional African practices that have the tendency of Africanising the post-colonial theory, in turn creating management academic disciplines and case studies for teaching purposes within Business Schools as well as practical everyday African interactions. The other path would be to universalise the post-colonial theory. Universalising it would mean marring it with like thinking global theories as attempted in this chapter while embracing African concepts like Ubuntu and "Noboa" as its cardinal anchors. The question then to ask would be if there exist or could there be an independent African theoretical model, one devoid of any influences?

Chapter 7

EDUCATING FOR SUSTAINABILITY

Concepts and Meanings of Education for Sustainability

Christopher Lim Gin Swee*[,‡] and Kim Beasy[†,§]

*Singapore Institute of Management, Singapore
†University of Tasmania, Tasmania, Australia

‡christopherlim@sim.edu.sg
§kim.beasy@utas.edu.au

7.1 A discursive history

Education for Sustainability (EfS) is a commonly cited term used in policy documents to denote a recognition for, and commitment to educate differently. Yet, discourses of sustainability have a complex and far-reaching history that influences the interpretations and subsequent implementation of EfS. Origins of sustainability discourses (by discourses we mean the way sustainability is understood, written about and practiced), lay in the deeply destabilising moments of the world wars, of social movements including the introduction of Eastern spirituality's' into Western modern societies and the environmental awareness of the 1950s and 1960s that was brought to life by the likes of Aldo Leopold and later by Rachel Carson in their scientific observations written with affectivity.

The complex discursive history of sustainability is relevant to understanding how EfS is positioned in contemporary discourses, as the interpretations and value-proposition of sustainability is by necessity, ever-changing. The following discussion is a brief sketch of the discursive history of education for sustainability.

During the 1950s to the late 1970s, problems of unsustainability were constructed discursively in a way unlike before. Many texts were challenging views of nature as "just" a supplier of raw materials, disconnected and separate to humanity. This time was characterised by a questioning of the modernist pursuit towards progress, as defined and achieved through economic growth.

Discourses of this time tended to peel back the layers of particular issues and think about what was at the core of these issues (Dryzek, 1997; Plumwood, 2002). For instance, issues of exploitation of resources was diagnosed as a problem of the economic model. Environmental degradation was diagnosed as a problem with the dominating relationship of humans over nature.

The 1950s to the 1970s can be characterised as a historical moment that sought to unravel the accepted cultural norms of modern societies and compelled questioning into what it means to be human in the world. The discursive setting predominantly conceptualised problems of unsustainability as cultural problems of alienation (Caradonna, 2014; Davison, 2001). Such as, Rachel Carson's "Silent Spring" that documented a powerful story of the toxic effects industrial society was having on the environment and consequently, on human health (Carson, 2002). Solutions proposed within these discourses reflected fundamental socio-cultural changes to the structuring of modern societies and interactions with the environment (Dryzek, 1997; Plumwood, 2002). Yet, these arguments were fading in the discourses by the end of the 1970s and new themes were beginning to emerge.

In 1987, concerns for environmental degradation, human impact and social justice were united under a common banner termed "sustainable development" and later "sustainability" by the Brundtland Commission, an arm of the United Nations. Since the release of the Brundtland definition, sustainability and sustainable development have become the dominant global discourse of ecological concern (Dryzek, 1997; Okereke, 2006; Torgerson, 1995). Soon after, education became common in international arenas as a way of delivering on "sustainable development".

The World Conservation Strategy was the first to redirect the goals of environmental education towards "education for sustainable development" (ESD) (Tilbury, 1995). It was determined that ESD would focus on the social, political and economic context of perceived environmental issues. This commitment was solidified by launching the World Conservation Strategy — Caring for the Earth: a strategy for sustainable living (The World Conservation Union, United Nations Environment Program, and World Wide Fund for Nature, 1991). The document firmly established education for sustainability as the central goal of environmental education in the 1990s (Tilbury, 1995). Education as paramount to achieving sustainable development was again highlighted in 1992 at the UN Conference on Environment and Development (UNCED): "Education [was identified] is critical for promoting sustainable development and improving the capacity of people to address environment and development issues" (United Nations, 1992, para. 36.3, p. 320).

In 2002, the UN General Assembly adopted a resolution that called for a Decade of Education for Sustainable Development (DESD 2005–2014). The discourse guiding the Decade was largely reformist and advocated for a shift in values, lifestyles and policy *within* prevailing forms of social structures and societies (Heckle and Wals, 2015). The "basic vision of the DESD [was] a world where everyone has the opportunity to benefit from education and learn the values, behaviour and lifestyles required for a sustainable future and for positive societal transformation" (UNESCO, 2005, p. 6). However, it is hard to find indications of comprehensive educational reforms driven by the Decade of ESD (Heckle and Wals, 2015; Sinnes and Eriksen, 2016).

With the close of the ESD Decade, a "new" policy context emerged (Payne, 2016). At the time of writing this book, imperatives such as the UNESCO Global Citizenship Education, 2014–2017 (GCED) and Global Goals for Sustainable Development, 2015 (GGSD) are driving transnational conversation and agendas, at least that is the intention. Yet arguably, there is little "new" to the sustainable development goals discourse. *Goal 8 — Decent work and economic growth*, emphasises the assumption embedded within the sustainable development goals that "sustainable development" is achievable within the current economic model.

Within the GGSD, Education for Sustainable Development remains the primary lens to which significant changes towards sustainability will occur. Evidenced by a target outlined in *goal 4–Quality Education*:

> By 2030, ensure that all learners acquire the knowledge and skills needed to promote sustainable development, including, among others, through *education for sustainable development* and sustainable life-styles, human rights, gender equality, promotion of a culture of peace and non-violence, global citizenship and appreciation of cultural diversity and of culture's contribution to sustainable development (emphasis added) (Project Everyone, n.d.).

It would seem that EfS is a discourse that no one wants to let go of in a hurry. And why would they? It provides a language that is accessible, open to interpretation and based on good intentions.

While we do not want to sound in any way pessimistic, the ability to change the ways that people *value, love, live and consume* in the world is no more likely to occur through EfS under this policy context, than the policy contexts of the preceding four decades of EfS and environmental education history. Why? Because emphasising education as a precursor to change does not acknowledge the complexity and necessity of an integrated approach to tackling the wicked environmental and social dilemmas of the 21st century. As noted by many scholars (Hill and Kumar, 2011; Orr, 2004; Peters, 2012), we must first be clear on what education is for.

If the fundamental purpose of education is to prepare people for, and build the capabilities required to live and work effectively in societies, then it seems at odds with this supposition that education can fundamentally change the society that it is directed to serve. And it is in this vein that we contend that EfS will be most successful when there is clarity in what the social and economic worlds of tomorrow will look like. Which of course is discussed elsewhere in the book.

What is more, we contend that education should be an integral and interconnected element of social systems. For the most part, most actors see EfS as primarily the concern of formal educators. While Tilbury (2007, p. 252) made this comment in the context of the ESD Decade, we contend it remains relevant today: "The need to position ESD, and thus the Decade, within frameworks relevant to all stakeholders across society is crucial".

In this way, we draw on discourses of industrial ecology and sustainable accounting as a means of reconceptualising and building an understanding of EfS that may be successful in penetrating into and connecting across diverse social and economic systems. We believe that EfS is most successful when there are EfS principles embedded in the social ecosystems of societies. We now turn to industrial ecology for an example of a discourse that promotes principles for sustainable development in industrial and corporate systems.

7.2 Lessons from industrial ecology

Industrial ecology is a systems approach that integrates economic and environmental systems (Biswas, 2012). This concept is relatively new, emerging from environmental management paradigms in the 1990s (Ehrenfeld and Gertler, 1997). This approach advocates a vision that industries are complex arrays of man-made ecosystems that exist alongside and necessarily connected and dependent upon natural ecosystems. Within industrial ecology, considerable dissipation of energy takes place as matter is transferred from one system to another. The emphasis of understanding is on recognising how connections to other systems can be made and on those connections that already exist (Erkman, 1997).

A key to understanding this concept is recognising that waste (that is a material, substance or by-product eliminated or discarded as no longer useful or required) does not conceptually exist. As noted by Jelinski *et al.* (1992, p. 793) "biological ecosystems have evolved over the long term to be almost completely cyclical in nature, with "resources" and "waste" being undefined, since waste to one component of the system represents resources to another". Industrial ecology takes this same principle and applies it to industrial ecosystems. Advocating that industrial ecosystems interconnect with and mimic the functionality of biological systems.

Below is a case study demonstrating a connection of ecosystems, working together and feeding off each other symbiotically.

7.3 Case study

Cervantes (2007) has drawn attention on the use of input-output diagrams at the School of Industrial Engineering of Terrassa, Barcelona to critique on the functionality of the industrial ecosystem in the city of Kalundborg, Denmark. This is with regards to the evolution of the Kalundborg industrial park ecosystem in 2001 (see Figure 7.1) where the modus operandi is for industry to use each other's by-products and share resources. At the heart of the collaboration network is the Asnaes Power Station, a coal-fired power plant. Steam from Asnaes is sent to Statoil Refinery in exchange for waste gas from refinery. Further, the removal of sulphur from its natural gas at Statoil Refinery enables it to sell the residue to Kemira, a sulfuric acid manufacturer.

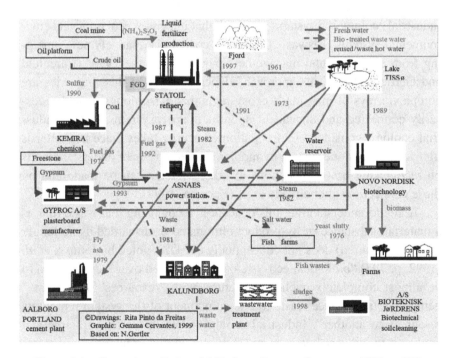

Figure 7.1.   Example on Industrial Ecology (*Source*: Cervantes, 2007, p. 135)

Asnaes creates electricity and stream from this gas and they are sold to Novo Nordisk, a pharmaceutical and enzyme manufacturer and for the heating of homes. The reuse of heat, in turn, lessens the thermal pollution discharged to a nearby fjord. Asnaes' fly ash is sent to a cement firm, Aalborg Portland. Gypsum from the desulphurisation process at Asnaes is sent to Gyproc for making gypsum wallboards. Within Kalundborg, fish farms use power from Asnaes and sell sludge from its ponds as fertiliser to nearby farms. Novo Nordisk gives away its own sludge, and businesses that recover contaminated effluents all play a collaborative role in the system.

We contend that industrial ecology offers a discourse to speak into the social systems that are currently largely disengaged from EfS discussions (Tilbury, 1995). Industrial ecology asks of its speakers to *think differently*. We find synergies in the objectives of EfS, with the principles of industrial ecology.

As many scholars have argued before us, we advocate that education for sustainability should be integrated into *all* aspects of social operations and societal structures. That EfS, ultimately, is about delivering messages or principles about how to live in the world and interact with the environments around us. We contend that following the educationalists model of delivering EfS has been largely ineffective as it emphasises education for youth. We believe that modelling EfS on the principles that have been shown as effective in social systems outside of education is a *better* model in which to be delivering education through. In this way, we suggest that industrial ecology offers an adaptable and broad approach in which to do so.

In the next section, we propose a model of EfS that we believe will be useful in constructing understandings of what it will take to transition towards a sustainability of the future based on the following principles:

- All systems (human and more-than-human) are connected to *other* systems,
- Waste of one system represents resources to another system,
- In the creation or maintenance of "systems", humans have a responsibility to relate with other systems in a way that does not inhibit any other system from flourishing.

7.4 Framework for educating towards sustainability

7.4.1 *Framework of knowledge-building*

The framework for education towards sustainability that we propose, positions knowledge building (KB) as a central tenet underpinned by a values framework based on the principles outlined above. For purposes of clarity, we view education as a lifelong process of learning, action and reflection involving all citizens. In this way, the model we are positing is intended to provide an overarching framework for thinking about education for sustainability across social systems including but not limited to formal educational settings.

There has been considerable research investigating the relationship between environmental knowledge, environmental attitudes and environmental behaviours (Ajzen and Fishbein, 1977, 1980; Burgess *et al.*, 1998; Cheng and Wu, 2015; Durr *et al.*, 2017; Hamari *et al.*, 2016; Newhouse, 1991; Rajecki, 1982). The one conclusion that can be drawn from the work in this area is that, human behaviour and decision-making is complicated.

Kollmuss and Agyeman (2002) published a comprehensive review of the internal and external factors contributing to people's performance of what is known in the literature as pro-environmental behaviours. Pro-environmental behaviours refer to behaviours that are recognised in dominant discourses of environmental education and sustainability to be less harmful on the environment; For example, recycling paper is considered pro-environmental in comparison to putting the paper into landfill. We draw on this research because this literature gives some insight into factors that may be more or less important in constructing an approach to education that is based upon the principles outlined in the previous section.

Relevant to the model we propose, Kollmuss and Agyeman (2002) found that:

- Institutional factors were relevant, in so far as particular behaviours could only be performed with the right infrastructure,
- Social and cultural factors were important in establishing the cultural norms guiding behaviours,

- Motivation towards performing pro-environmental behaviours are largely overridden by more immediate motivating factors,
- Environmental knowledge is not a pre-requisite for pro-environmental behaviours,
- Values are important in shaping intrinsic motivation.

Kollmuss and Agyeman's (2002) study highlight the importance of social and cultural norms in developing pro-environmental behaviours. In the model we have developed, constructing knowledge is a central tenet because of the need to build cultural norms in societies. The delivery of information and building knowledge is one way of contributing to and establishing culture. We contend that societies need to be educating *differently* with *different* principles directing curriculum and pedagogic and andragogic decisions. In this way, information that is "taught" (or constructed) and the way of delivering (or constructing), is paramount to its receipt and subsequent interpretation. All of which influences how people position themselves in the social systems that they are embedded.

And in this way, we see that implicitly *and explicitly* embedding the principles we referred to earlier which we conceptualise through a values framework, offers potential in spaces for educating towards sustainability. As noted by Kollmuss and Agyeman (2002), values are shown to drive intrinsic motivation.

We propose a KB EfS framework for studying education towards sustainability: see Figure 7.2. We have drawn on complexity theory to inform the development of the framework. The KB framework proposed uses complexity learning theory to conceptualise how meaningful insights into unfamiliar and dynamic situations can be achieved (Scardamalia and Bereiter, 2006). We have drawn on this theory in recognition that problems of unsustainability that exist in the 21st century are complex and dynamic and need *complex and dynamic* solutions.

The complexity learning approach encourages students to expand their world-views beyond boxed-up, linear learning. The pedagogical implications are in rejecting "defensive reasoning and defensive routines" (Argyris, 2004). As shown in the *y*-axis of box diagram of Figure 7.2, following the framework encourages double-loop learning (Argyris, 1993).

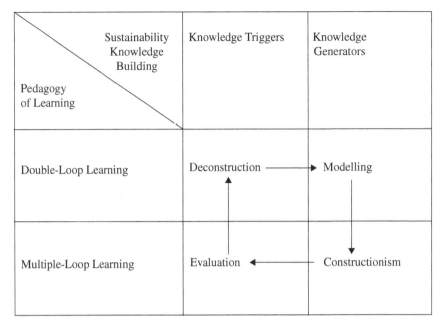

Figure 7.2. Avenues of KB in Sustainability Education

This affords greater iteration in processes of learning while multiple learning loops are occurring within and without the "problem space".

Double-loop learning incorporates theory-in-action through the iterative processes embedded in action research which has a valuational and moral component that motivates the KB learner and the world to become developmentally better (Argyris and Schon, 1978; Nielsen, 2016). The use of double-loop learning in KB "problem space" is in contrast to single-loop learning. Single-loop learning is linear in design, generally moving from a state of unknowing to presumably "knowing". In other words, single-loop learning does not incorporate robust processes of reflexion into learning.

Multiple-loop learning is even more iterative in knowledge construction and actualisation. It is a suitable pedagogy for our contemporary VUCA (volatile, uncertain, complex, and ambiguous) world. The latter is a military acronym which was introduced by the U.S. Army War College in the late 1990s (Chadha, 2017). The VUCA world is disruptive and challenges the relations of power between students and teachers. In this process, students become partners in the unfolding learning journey. Multiple-loop learning incorporates processes of construction and evaluation and takes learning from deconstruction and modelling into a place of real-world application. By necessity, multiple-loop learning can encourage autonomous and differentiated student learning in EfS.

The idea of KB is to help students' reject certainty thinking and to *think differently*. In doing so, they increase their capacity to create knowledge and other ways of being-in-the-world. The proposed framework, as shown in the *x*-axis is inspired by Cummings and Angwin (2015, p. 164) who regard "encouraging strategic initiatives" as being based on multiple creativities (knowledge triggers) and discoveries (knowledge generators). The cross-section of KB concepts between the *y*- and *x*-axes in turn gives rise to four avenues of sustainability educational practices: deconstruction, modelling, constructionism and evaluation.

7.4.2 *Deconstruction as EfS*

Deconstruction involves breaking down pre-existing knowledge that learners have about how the world is constructed. The purpose of deconstruction is to create opportunities for thinking otherwise. Teachers are positioned as facilitators and at the same time, provocateurs in presenting materials that create moments to afford students opportunities to deconstruct worldviews and think otherwise. Provocations should be introduced in a way that is age appropriate and sensitive to the needs and abilities of the learners.

A pedagogy of deconstruction encourages learners to think outside of the box on problems relevant to questions of sustainability. There should be flexibility in the delivery of learning material to enable student-directed conversation, discussion and exploration to challenge and "test" their preconceived ideas related to the issue. The intention is to create disruptive changes in how students think about the world.

There are a number of strategies that teachers can use in creating moments that afford students opportunities to deconstruct knowledge. The EfS framework proposes knowledge triggers as a means to help learners "poke" and find new insights. Snow and Benford (1988) distinguished three types of framing knowledge triggers — diagnostic, prognostic and motivational.

Diagnostic framing involves first identifying a problem, then working to assign responsibility of its creation. The intention of using diagnostic framing in an educational context is to encourage students to think about the complexity of problems of sustainability and think about how they came to be; beyond problem identification. Diagnostic thinking can only be used in so far as encouraging depth of thinking and thought-extension. Problems of sustainability are complex and often cannot be attributed to any one entity or agent. Further still, framing problems of unsustainability are always dependent on the perspective of who participates in the act of framing, including both student and teacher. In this way, we suggest educators think of diagnostic framing as a means of beginning a process of deconstruction, rather than a path to its achievement.

Prognostic framing is positioned pedagogically as solutions focused. In this way, students are asked to brainstorm or create solutions to identified problems. Students engage in processes of deconstructing problems of unsustainability through imagining otherwise. A teacher's role here may be to provoke and encourage critical and imaginative responses. For example, a teacher may present the problem of paper waste of a classroom and ask students to come up with solutions. Students could then brainstorm possible solutions to this problem. While this is a small scale example, the same framing could be applied to global issues such as poverty or deforestation. Students should be encouraged to think about solutions that are both possible and perceived as impossible. This way, students engage creatively with problems and challenge their ways of knowing. Teachers should in these moments, be encouraging of student's creativity and challenging conceptions of impossibility.

Motivational framing calls for society and social movements to trigger action. In this way, students deconstruct particular knowledge from a value-based positioning. While the above two approaches engage students in rational and reason-based logic (both challenging and at times promoting this form of thinking), motivational framing encourages problems to be ethically and/or morally located. For example, learning material on farming may be presented from a position that farmed produce should be traded fairly to meet the needs of farmers. Positioning problems of unsustainability morally or ethically has previously been contested and relative to the cultural the social context. However, we propose that motivational framing can be based on the principles proposed above and at the same time, reflect particular learning contexts.

For example, we suggest that there are opportunities for deconstruction to occur when students frame and reconstruct through an inquiry. Through photo-voice narrative, students took photos and used photo and voice storytelling to trigger conversations and critique on community issues in a Costa Rica rural mountain town. The authors, Cook *et al.*, (2016, p. 58) cited areas of deconstruction in the students' learning of coffee cultivation through the following student recollection:

The more trees, the more shade and better water distribution from the rain. But with coffee being higher up, it is colder and harder to grow other fruits. It is hard to grow banana trees here, because they need shade. But, most of the fruit trees here are not to produce fruit, but to protect the coffee plants from the sun and rains.

In another article, Madden and Dell'Angelo (2016) showcased the use of photo artefacts in the form of photo-journals in a blended online environmental science course to portray student impressions on "living bio-regionally". That is, one where students are encouraged to explore their relationships in their geographic place. These entries such as those exemplified in Figure 7.3, demonstrate how students reflected on the relationships they had with the physical environment and began to deconstruct and challenge these relationships that had previously been implicitly accepted.

7.4.3 *Modelling as EfS*

The importance of experiential learning has been evidenced in the literature for a number of years (Boud *et al.*, 1993; Kolb, 2014). In alignment with this research, we see modelling through experiential learning as fundamental to education for sustainability. In this way, modelling is not intended to denote social learning, that is, through the modelling of behaviour. Rather, we position modelling as engaging in practical processes of

Living Bioregionally

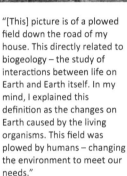

"[This] picture is of a plowed field down the road of my house. This directly related to biogeology – the study of interactions between life on Earth and Earth itself. In my mind, I explained this definition as the changes on Earth caused by the living organisms. This field was plowed by humans – changing the environment to meet our needs."
Participant 3

"As humans, I think we focus on the ways we shape the world around us, just as our footprints leave markings on the sand. We see our impressions on the land around us fairly easily. Sometimes we forget the power of our environment, because in one swift pull of the tide all traces of our footprints have been reduced to minor ponds of collected salt water or disappeared altogether."
Participant 4

Figure 7.3.   An Avenue of Deconstruction (*Source*: Madden and Dell'Angelo 2016, p. 28)

building and/or conceptualising frameworks or prototypes in processes of addressing problems of unsustainability. These prototypes are espoused representations of reality and give the learner playful opportunities to try, create and engage in processes of problem solving.

Examples of modelling already exist in many schools. For example, school gardens can be harnessed for the opportunities they present for students in designing, trialling and applying different methods of food growing. Such models may provide a fertile soil for students to exercise their cognitive flexibility and discover various problems and solutions that can be applied to problems of unsustainability.

While Wolsey (2015) identified three tiers of learning, we suggest that EfS must engage across these tiers in complex and messy ways.

Wolsey (2015) identified one tier of learning as observational which may include students observing teacher practice and instruction in the garden. In tier 2, learning traditionally may have included students being actively engaged in a specific curriculum such as an Edible School Yard project on how to work in the garden and kitchen. Through this form of engagement, students may learn how to construct raised beds, calculate optimal space for plant growth, etc. In tier 3 of learning, students may use the garden as a means of connecting to worldly issues, such as obesity and diabetes and work on ways of promoting eating healthy foods. Examples of this model are evidenced in the Chinese-style university eco-garden, in the emphasis of promoting an awareness of harmony between nature and humans. Through an eco-garden design, students become aware of the spontaneity, simplicity and naturalistic cosmology that so determines Daoist and neo-Confucian philosophy (Cheang *et al.*, 2017). Each of these tiers offers discrete learning opportunities that can work towards education for sustainability. These learning opportunities reflect the need in Daoism and neo-Confucianism to reflect a KB that embodies a harmonious, yin-yang co-existence between man and environment. That co-existence encompasses the *Dao* (Way) of *Tian* (Heaven), which is the natural order of nature (see Gerstner, 2011; Li *et al.*, 2016). If there was continuous incongruence between human activities and the environment, unsustainability or even calamity may befall the natural order of things (Lao-tsu, 1991).

The fact is that we find increased learning opportunities when the three tiers of learning have been integrated into a holistic and interconnected process of KB. For example, school lunch programmes in Japan are delivered under the guise of the "School Lunch Act" enacted in 1954 and revised in 2008. The School Lunch Act legally requires schools to provide meals to students during the day and since 2008, requires schools to promote *Shokuiku* (Tanaka and Miyoshi, 2012). This denotes the "as acquisition of knowledge about food as well as the ability to make appropriate food choices" (Ministry of Agriculture, Forestry and Fisheries, 2006). An example of a school lunch programme in an elementary school in Saitama has integrated the growing, harvesting, processing, cooking and consuming of lunches into the curriculum and school day (CafCu Media, 2015). Students are actively contributing to making nutritious meals from locally sourced produce. The learning that occurs in this and similarly integrated programmes extends beyond a tiered system of learning toward a cyclical, multiple-looped process of KB about food systems (Tanaka and Miyoshi, 2012).

We propose that in these programmes, children are engaged in modelling food systems that reflect the principles of an education for sustainability. As reported by Tanaka and Miyoshi (2012), benefits of the programmes include increased awareness about diet and nutrition and improved quality of life of children. Further to this, incorporating lunch preparation that is connected to kitchen gardens, models a process of food production that challenges and re-imagines how food systems are constructed in many societies globally.

7.4.4 *Constructionism for EfS*

Processes and activities of modelling follow onto processes of construction through the engagement of students in systems of societies. In such a context, much action learning may take place (see Revans, 1981), and augment an:

> "insight into the posing of questions by the simple device of setting them to tackle real problems that have so far defied solution. We may structure our argument from the outset by identifying the acquisition of programmed knowledge as P, and of questioning insight as Q, so writing the Learning Equation as: $L = P + Q$. In this, our principal interest is in Q, the idea of action learning. We do NOT reject P; it is the stuff of traditional instruction." (Revans, 2011, pp. 2–3).

In action learning, the loops of learning tend to be further multiplied. Rather than following a fixed agenda, the posing of penetrating questions (Q) from an origin of traditional instruction or programmed knowledge (P) in educational institutions may precipitate constructionist approaches to confronting issues on sustainability. In this way, we position educational institutions as a training ground and a connective element to the social fabrics of societies. This is in contrast to traditional conceptualisations of schools and educational institutions as producing skilled citizens *for* society. Following the example of the Japanese school lunch programme, a construction onto this programme may be in the design and trialling of programmes in other social institutions, such as, work places or office blocks.

Constructing actively engages students in real-world systems through the school environment. Students learning from a constructionist perspective become active participants to construct their own meaning, resources and innovative solutions in society to confront the challenges posed by a situation. It is possible then that this constructionism would involve "double transfers" of learning from nature to society and back again from society to nature. The following case study of the Great Barrier Reef exemplifies multiple constructions undertaken by educational institutions. In this way, educational institutions are engaged with industry and government to co-create solutions for problems of unsustainability.

For example, Reef Warrior scientists are helping to prevent further deterioration of the Great Barrier Reef in Australia (Smith, 2017). The Great Barrier Reef is a significant natural emblem for sustainability as it was the first coral reef ecosystem to receive World Heritage recognition as a sustainable resource. However, as a result of increased pollution, the crown-of-thorns starfish with its appetite for coral polyps has multiplied within the reef ecosystem. Scientists recognised the need to reduce and control the starfish population. At first, divers were used to inject the starfish with a toxin. However, due to increased population outbreaks, this method soon became ineffective. Scientists from the Queensland University of Technology constructed a submarine-like robot to patrol the seafloor. When it encounters a starfish, an injector arm is released to inject the starfish with toxin. Collaborative relationships between education institutions, industry and government provide increased opportunities for EfS learning through constructions.

A further example is the work of scientists from Griffith University to monitor acidity levels of the reef waters. Acidity levels can affect the survival and "flourishment" of marine life on the reef. When sea temperatures become too warm, the symbiotic relationship between the corals and algae become strained. And so scientists worked to find algae, known as crustose coralline algae, to act as a bio-indicator of increased acidification. Further still are examples of scientists from the University of Sydney who are using robotics and stereo imaging to develop 3D Australian reef maps for studying coral bleaching phenomenon.

The examples cited above demonstrate how educational institutions can be contributing to solutions for problems of unsustainability. We contend that engaging secondary and primary school students in similar partnerships has the potential to inspire young people and encourage their authentic engagement in problems of unsustainability.

7.4.5 *Evaluation as EfS*

Thinking differently demands that thoughts and actions are *critically* evaluated during processes of KB. Evaluation by definition is the making of a judgement according to some method of assessment (Oxford University Press, 2017). People are continually evaluating information throughout a day, consciously and unconsciously through cognitive processing. We acknowledge that there are multiple ways of undertaking evaluations and that there is no one *correct* methodology. Each evaluative act must be context specific and sensitive to the social, cultural, political and environmental conditions. What we do propose is that education for sustainability demands undertaking evaluations as an integral part of building knowledge, and that these evaluative processes *are* critically and sensitively undertaken.

We contend that incorporating critical reflection into learning activities can support students in evaluating their learning. Reflection was first coined by Dewey as "an active, persistent, and careful consideration of any belief or supposed form of knowledge in light of the grounds supporting it and future conclusions to which it tends" (1933, p. 6). Critical reflection is the highest level of reflectivity and involves thinking *and* problem solving (Yost, Sentner and Forlenza-Bailey, 2000; Bonney and Sternberg, 2011). It moves beyond the consideration itself, for example a learning activity, to consider the assumptions and context that the consideration is embedded within. For example, what assumptions about the world were evident within the learning activity?

An evaluation of the sustainability learning experience is important both for learners and educators in evaluating learnings. A component of initial educator training is in learning skills to be a reflective practitioner (Calderhead, 1989; Gore and Zeichner, 1991). In education for sustainability, we see the need for *critical* reflection embedded in teacher's evaluative frames for promoting and transitioning towards sustainability.

Concepts of critical thinking, reflection and critical reflection are important in performing rigorous evaluations. This is evident in educational settings (Brookfield, 1995; Calderhead, 1989; Gore and Zeichner, 1991), and beyond (Luederitz *et al.*, 2016). However, any form of evaluation, whether it is applied to formative, ex-ante evaluation or ex-post evaluation of completed experiments, is based upon a set of criteria. In some cases, this criteria may be based on an individual's moral and/or sense of ethic, in others, objective performance indicators may be defined. In any case, we propose that evaluations of education for sustainability programmes, should be made in alignment with the principles proposed earlier in this chapter.

The University of South Pacific, for example, has led and used integrated ESD objectives (see Figure 7.4) to chart forward a philosophy-based *Pacific Way* for sustainability evaluation in higher education (Corcoran and Koshy, 2010). The Pacific Way is based on a formative spiritual imperative that pervades Oceania.

"In Fiji, it is called *Vanua*. It includes the heavens, Earth, and the underworld or afterlife. It includes all time — which is not linear but cyclical... Often, death is associated with misuses of land, so one learns to respect it early in life and to know it as the foundation of education and sustainability." (Corcoran and Koshy, 2010, p. 131).

The Pacific Way has resulted in a reflective consensus-building on sustainability by various stakeholders of the region. This has been based on the conceptualisation of *Vunua*. The concept has multidimensional meanings and implications. It is a term for land and place or environment. But it also recognises understandings of land as encompassing people, their philosophy of living and society, and their need for survival in the economy. Hence, environment, society and economy are shown in a holistic embrace in the left side of Figure 7.4. Further, *Vanua*, it may be argued, represents the spiritual bond between people and nature. Hence,

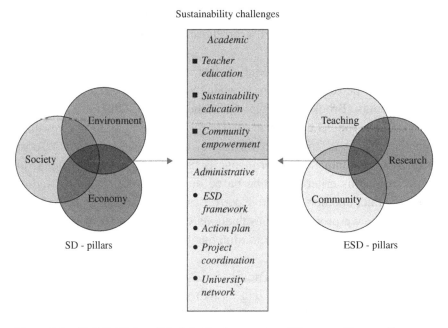

Figure 7.4.   Holistic Vision of Society, Environment and Economy (*Source*: Corcoran and Koshy 2010, p. 136)

Vanua has justification to support the related ESD pillars of teaching, research and community involvement. This is depicted in the right side of Figure 7.4. At the centre of the diagram, are the academic and administrative challenges of sustainability education. On the academic front are challenges for (a) teacher education including building capacity in Pacific arts and culture; (b) sustainability education as in sustainable islands and oceans development and ESD resource materials development; and (c) community empowerment such as the teaching of environmental stewardship for young people. On the administrative front, establishment of (a) an ESD framework that is to be led by a board, advisory committee, project leader and coordinators; (b) an action plan for academic programme development and delivery; (c) project coordination of various action research projects in various thematic areas; and (d) a university network that links various universities to the common purpose of forging education for sustainability. There is no doubt that multiple evaluations and reflections may be made based on the framework of Pacific Way, which in the last analysis, acknowledges all humans are one in all the things we do and spending time with nature helps humans appreciate this unity.

Besides the University of South Pacific initiative, various other universities and observers have made declarations in favour of, more formal summative sustainability audits. For example, Lambrechts' (2015) study on the application of AISHE (Auditing Instrument for Sustainability in Higher Education) in a Belgium university. The audits and calls to be in empathy with the purposes of sustainability, however, are not a straightforward endeavour. Rather, according a reflection of a third-year student of a teacher education programme in a Baltic university it is a journey to be connected to nature that would need time and resilience:

> "Someday, I would like to live in connection with nature. I think the more we consume, the worse and I am really in favour of everything [that protects nature]... but I think that we have not yet come to the situation where people and nature can unite...the world has to change for that. I believe that these changes will occur; just a lot of time has to pass, maybe 200 or 300 years." (Raus and Falkenberg, 2014, p. 110).

In this way, the pre-service teacher recognises the importance, yet the patience required in transitions toward sustainability.

7.5 Teachers: Bridging the gaps between self and world

Education for sustainability scholars purport that educators need to be teaching from an ecologically aware position. This means that as educators, there is a need to reflect upon our own positioning in the broader ecosystems and chains of connections that we are embedded. The disconnection between people from nature has contributed to the ecological crisis that marks the 21st century (Franklin, 2002). In many ways, cultural values that undermine the connection between people and planet have endorsed the domination of nature (Naess, 1995). We contend that in educational contexts, teachers should embrace connections with nature, teach through the ecological-self and the identities that are connected to the natural environments that they are located in.

Tasmania, the island state of Australia, has been a fertile ground for thinking about the implementation and take up of education for sustainability across school based sectors. The curriculum context within Australia, means that Tasmanian educators must teach in align with a national curriculum and early years framework. Educators are supported in the delivery of sustainability based curriculum through the location of "sustainability" as a cross-curriculum priority in the Australian curriculum. What this means is that educators teaching across grade levels are expected to be linking learning experiences from across the curriculum to sustainability (Australian Curriculum, Assessment and Reporting Authority, 2016) and educators teaching children who are under five years of age must teach in align with the Australian early years learning framework (Australian Government, 2017). An approach to support teachers in doing this has been engaging educators through a number of professional development opportunities.

A number of professional development sessions were conducted with early childhood educators specifically on implementing education for sustainability in Tasmania (Dyment *et al.*, 2014). The facilitators found that participation in the sessions extended educators' understandings of sustainability from predominantly an environmental perspective before participation, to including social dimensions of sustainability at the end of the sessions (Dyment *et al.*, 2014). The learning experiences that early childhood educators reported implementing in their classrooms predominantly focused on the environmental/natural dimension of sustainability, however, the authors conclude that "[these learning experiences] hold significant promise for sowing seeds of transformative sustainability learning and children's active citizenship" (Hill *et al.*, 2015, p. 20). This suggests that having a top-down approach, that is embedding EfS in policy and curriculum objectives, must be complemented with support to educators on the ground.

What this section has sought to demonstrate is that implementation of education for sustainability must occur as a multipronged approach. In the Tasmanian context, there remains high level support through policy and curriculum documents for implementing sustainability across learning areas, yet, there remains barriers in teacher's knowledge and skills in knowing *how* to "do" EfS (Dyment *et al.*, 2015;

Dyment *et al.*, 2014; Hill *et al.*, 2014). In this way, we contend that affording teachers' time and space, as was done in Uganda, to develop their own practice of sustainability is necessary in successfully teaching for sustainability.

7.5.1 *Learners: Motivating learners in EfS*

Ecojustice education is an emerging field of theory and inquiry that focuses on the cultural crisis of the environmental crisis. These scholars are concerned with the cultural practices that are shaped by the ideology underlying the industrial revolution. Hyperconsumerism and the domination of nature and all other species typify this mind-set and its accompanying practices. The cultural crisis includes an absence of regard for the ecology of all of life. Ecojustice educators maintain that the environmental crisis is a cultural crisis — and, as such, requires people to think and behave in ways that are less detrimental to the sustainability of life and natural ecology of earth. Education for sustainability should endeavour to engage learners in learning that is critical, challenging and encourages reflexion about how people live in the world. Learning activities that align with what students like, or are interested in, increases the likelihood of deeper engagement with the material/activity (Clarke *et al.*, 2001; Farias and Fairfield, 2010; Taylor *et al.*, 2004). While educators have agency in course material, it is students who are actively choosing to not/engage with it (Eastman *et al.*, 2011). In recognition of the relationship between students' motivation or willingness to learn and actual learning, we contend that education for sustainability should be pitched in ways that are contextually relevant and reflective of the learners' interests (Cole *et al.*, 2004; Karns, 2005; Zocco, 2009).

In this section we go on to describe a course delivered to adult learners about sustainability and educating for it. This example draws on the experience of one of the authors, who delivered the course (spanning 13 weeks) to second year pre-service teachers as a component of a Bachelor of Education in Tasmania, Australia. The unit received very good feedback from learners. One learner noting, "this unit made me see our world from a different perspective...". We detail the structure of the course as well as list some of the materials and activities that were used in it. This course was designed to the specific context of the learners, taking into consideration their prior degree learning. It was also located in the social, cultural and political context at the time of offering.

The course was delivered using a module structure. The first module engaged learners in thinking about their worlds. They were asked to "deconstruct" how they understood the world and to think critically about global systems; social and ecological and the interconnections within them. Materials drawn on in this module included Johan Rockstrom's "let the environment guide our development" TEDTalk, Jacobs (1999) seminal piece on sustainable development among others. Learning tools such as ecological footprint calculators, were used in this module to challenge learner perceptions about their positioning in the world and the systems that they are a part of.

Module 2 presented learners with various ways of "doing" sustainability education (in this context, the course was delivered to pre-service teachers). Content included thinking with sustainability education pedagogies (Cotton and Winter, 2010; Killen, 2013) and unpacking concepts of agency and "action" (Jensen and Schnack, 1997). Throughout this module, learners were engaged in "modelling as EfS" through experiential learning of the concepts presented. The learning activities mimicked the realities of their future environments (such as their own classroom and students) and gave the learners playful opportunities to try, create and engage in processes of problem solving of and with the approaches.

Module 3 aligned with the concept of "constructing for EfS" through actively engaging students in real-world systems through the school environment. Learners were asked to think about real-world systems and how they could be engaging with them personally, as well as thinking about how they could be incorporating actions to address problems of unsustainability in their future classrooms. Materials utilised in this module included stories, blogs and resources documenting the constructed solutions of others as stimulus for thinking about their own applications to problems. For example, the idea of "sustainable consumption" was presented and resources to explore this included links to "PlasticFreeJuly" and materials related to "zero waste living".

The assessment of the course focused heavily on learners' development of reflection and reflexivity. Assessment addressed questions of the learners' ability to think critically about problems of unsustainability, as well as their ability to reflexively construct and implement solutions to address them. In this way, "evaluation as EfS" was a core component of the course in design and delivery.

The design and delivery of any educative programme should be thoughtfully and reflexively compiled. Education for sustainability programmes are no exception. We include the example above to highlight the structuring of a course, designed to speak *to* learner context and speak *back* to dominant ideas about nature–human relations and roles of people and educators.

7.6 Concluding remarks

This chapter began with the presentation of a history of sustainability education. We considered the challenges of discursively positioning education for sustainability and how this has changed and continues to change over time. Sustainability education, like sustainability, remains a contested space in which to think and work. The EfS framework has opened scope for society to foster KB oriented schools, universities and organisations without walls to study and practice our responsibilities *on sustainability*.

The nature of sustainability brings into questions fundamental beliefs and assumptions about the world and our positioning as humans within it. We believe that education for sustainability is primarily about challenging these beliefs and assumptions and moving forward to think about how things could be otherwise. This belief is reflective in the principles that we contend should guide education (ambitiously), specifically, education for sustainability. Our vision would be a time and space when sustainability education becomes obsolete. It is no longer required because ways of being-in-the-world (our social and economic systems), are constructed around concepts of sustainability. As we move along that trajectory towards independence, however, that metaphysical state of being-in-the-world would be one where the scope of learning is multidisciplinary, inclusive and longitudinal and as transformational as EfS KB can be.

Chapter 8

EMERGING BUSINESS VALUES

8A — Changing Values

Nick Birkinshaw

Ecus Ltd, Sheffield, S61 2DW, UK

nick.birkinshaw@ecusltd.co.uk

Ecus Ltd. is a multidisciplinary environmental consultancy with offices across the UK. We are contributing to the Intrinsic Capability project to help identify the inherent connections people have to the environment and how these can be used by organisations and business to better understand and face their customers.

This contribution on "Changing Values" aims to discuss the response of organisations and business to the intrinsic value of the environment to their customers as well as highlighting opportunities for collaborative working in allied organisations and innovation in business practice.

8A.1 Development of the sector personal observations/experience

My personal experience of ecology started when my father, a teacher, returned from a teaching conference in the early 1980s where the subject of Ecology as a growth study area for Universities was being discussed. He correctly predicted this would be a great subject for me to pursue. When I looked to study for a degree, Ecology as a dedicated topic was relatively new and only a handful of courses were available. I studied at the University of Leeds and on completing the course looked to find a career as an Ecologist.

I recall a careers department that did not really know what the subject was and could not guide me other than to working as a countryside ranger in a national park or local authority.

I took the ranger route but also worked in retail which sparked my interest in business. Whilst working for the local authority as a seasonal countryside ranger a colleague was developing a business as an Ecological Consultant and suggested a further degree at University of Bristol that was aimed at Environmental Professionals and particularly Environmental Consultancy. To my knowledge this course in "Ecology and Management of the Natural Environment" was one of only a couple in the UK at the time — 1996.

At that time the environment market was emerging and comprised specialist departments in large engineering companies, some developing university spinouts as well as some independent start-ups formed by individuals. My business Ecus Ltd. first opened its doors in 1986 as a spin out from the University of Sheffield and was a pioneer in the sector.

As a professional ecologist at the time you were generally dismissed as getting in the way or seen as a hindrance by established engineers and project managers but the younger generation of engineering graduating from universities generally adopted and integrated environmental requirements without question — it was assumed that the environment was part of the job and on the hole the attitudes changed across all generations as a result.

Up to 2008 financial crisis larger companies were busy with buy and build models adopting environmental specialism into their business. As the recession hit there was a tendency for these larger companies to drop environment staff as they returned to playing to their strengths in proven and disciplines.

Ten years forward and businesses at all levels that traded through the financial crisis are typically more resilient with "built-in" disaster plans but they are also working more collaboratively which makes them more agile and more able to change to differing market conditions — for example they tend to be less dependence on one client or sector.

The larger companies retain environment staff but have less expansion in this discipline in their business against a backdrop of collaborative working with Small-to-Medium Enterprises (SMEs) to cover resourcing on larger projects driven by contractually required by large infra structure clients.

Typical environmental businesses in current UK market can be very broadly viewed in three types:

- Large multinational multidisciplinary companies — facing developments in particular infrastructure projects as "tier 1's suppliers". Growth driven largely by acquisition as organic growth is too slow to give sufficient return.
- SMEs — typically covering smaller planning driven projects, supporting utilities with specialist consultancy in specific sectors and supporting tier 1s with discipline specialism.
- Sole trader/small specialist business' formed around a key individual usually highly specialised niche discipline provider.

8A.2 Development of environmental expectation and the episteme shift

The move of people from countryside to city in industrial revolution saw for the first time the majority of the population losing contact with the natural environment. In the drive for industrial growth the environment was at best a low priority and this continued through the early part of the 20th century exacerbated by the two world wars.

Other sciences require a high level of understanding but environmental affects everyone with a high level of interest and one can see and identify with the impacts and changes to the environment in, for example, industrialisation. This engenders a sense of ownership in the environment from the public that other sciences do not generally trigger.

Post World War II economic expansion was coupled with population growth and the emergence environmental awareness. This was picked up on by the baby boomer generation some of whom were noted for rejecting "traditional values" and including popularising environmental issues.

Whilst still outside the mainstream the developing values of environmental awareness and sustainability were entering education and influencing the emerging Generation X, myself included. This helped position these values in the mainstream of current thinking, policy development and business operation to the extent that the younger generations of Millennials and Generation Z simply expect and require what was once a fringe value to be present in daily life.

This has significant implications for business on two fronts. Firstly, a company typically needs to have an environmental awareness of its operations and be communicating that to its customers as part of its marketing and many are doing this in order to gain a sales advantage over competitors. Secondly, it is essential for companies to be an attractive employer and that means traditional incentives of salary, prospects and status are not enough. Increasingly employee priorities are shifting to work life balance, organisation quality and ethos including environmental responsibility.

This is certainly my experience in the environmental consultancy market and in a traditional P&L account these are issues that drive the top line figure and which supports with the Triple Top Line affect described by McDonough and Braungart (2002).

People have an intrinsic interest in the nature, even when they might not be particularly exposed to it. It is easy to see for example how aboriginal people are intrinsically linked to the natural world and see themselves as part of it. The links to the concrete jungle of the modern metropolis are often less obvious — but they are there.

In 2017 David Attenborough's Blue Plant II attracted 14 million viewers for its opening episode — roughly the population of Somalia and about 21% of the UK population. Similarly the popular rural magazine programme Country File attracted 9.5 million viewers to one episode in 2016 — roughly the population of Hungary. Clearly this is a crude metric but there are lots of distractions in a modern world, lots of other things people could be doing with their time and for a TV programme to be capturing the attention of over 20% of a country's population is worthy of note.

It would appear to me at least that these people are connecting with the natural world albeit from the comfort of an armchair and these numbers are of interest to any business trying to gain competitive market advantage. Cynical? Maybe but money is like water and will find a route around pretty much anything that stands in its way, including traditional accounting techniques. As analytics improve and attitudes change, market forces are shifting to align business with the human intrinsic connection to the natural world.

8A.3 How businesses (and not politicians) are recognising and facing the shift

Modern science documents our world as a myriad of complex relationships and the narcissistic neo-classical view of the world is being replaced by a more outward looking and collaborative view that draws from our intrinsic connection to the natural world creating a new "Primal Episteme" that places environmental consideration at the centre of peoples thinking.

Whilst Politicians still largely appear to be fighting the neoclassical battle progressive businesses are starting to recognise the new environmental and social values and are responding to the changes within the new consumer demographic.

There is a tendency for traditional "bulldozer businesses" to push against legislation and policy drivers to try and remove the obstacle to profitability and we still see today in the lobbying of politicians to remove what some individual companies see as unnecessary or excessive legislation. In these cases the environment is seen as "just another expense". However, often business sits above politics and make decision of their own volition driven by client demands.

Naturally these decisions are still derived from a desire to increase market share and boost shareholder value but companies will deliver what their customers want in order to achieve this and so environment is often a catalyst for decision making and not just finance.

International business leaders are starting to recognise and face changing attitudes towards environmental issues in order to protect and add value to their business. For example, Unilever have criticised UK policy on renewable energy saying the Governments position does not reflect their customers' desire for increased levels of energy provision to come from renewable sources. Paul Polman CEO of Unilever said in a 2017 interview with the BBC "Ultimately if any of us won't act, we put the lives of many people at risk", and continued "What we're talking here about is the future of humanity to some extent. This concept of man's dominance over nature is rapidly being rewritten." (Polman, 2017).

Michael Bloomburg in 2017 interview challenged the withdrawal of the USA from the Paris Climate Accord and noted that withdrawal didn't matter as local and government and business were working towards hitting the targets anyway — "Cities took action, they painted roofs white to reflect heat, converted to energy-efficient light bulbs and cars. They stored water to reduce runoff". Bloomberg says that companies want to be environmentally friendly so they can attract the best talent, and are under pressure from customers and stockholders to be eco-friendly. He has said "We've closed coal powered plants at the same rate since Trump was elected as before, he said. The reason is that plant owners are influenced by the economics — natural gas is cheaper and cleaner than coal — as well as what other people or their families think about living near a coal-fired plant even if it's not the American government, it's the American people that are behind this." (Bloomburg, 2017).

I have no doubt we could pick apart the drivers behind any of these statements but what is clear is that the environment and sustainability is rising to the top of many CEO's agenda's in order to future proof business to face changing public attitudes and attract the talent needed to drive the sales in a changing social environment.

Environmental issues are often referenced within a company Corporate Social Responsibility statement (CSR). Implementation of CSR commitments typically start with measures like cutting carbon emissions or purchasing proportion of energy from renewable sources. However, there are also opportunities for business' to be outward looking, for linked up collaborative working with like minded organisations to enhance existing resources.

As the environment becomes more mainstream the subject diversifies and develops new layers of complexity. This is resulting in an increased level of collaborative working between organisations, particularly at the project level on large scale infrastructure schemes such as High Speed (HS2) or highways schemes. The diversity of organisations involved limits risk to the project offering project stability, it is inclusive, benefits regional and local scale companies and economies and captures and integrates specialist skills promoting innovation. This is showing a shift from the triple bottom line to the triple top line business behaviour driven in part by environmental considerations.

The challenges in project collaboration are many but typically include tier one companies needing to get SMEs familiar with the systems and processes needed for a large organisation to operate and conversely the SME needs to be make the tier 1 managers understand the lifeblood of cash-flow, the cost of systems up-scaling and the importance of forward order-book forecasting and visibility.

There are a number organisations whose business is driven by the environment or that have an increasing awareness of the importance of environment to their continued success and these do not necessarily fit the typical image of a large successful corporation. The National Trust, The Royal Society for the Protection of Birds (RSPB), Crown Estates, the Forestry Commission and various Water Companies and Utilities with large land holdings are all examples of successful businesses that rely upon or market the environment as a resource or "product".

They are often linked by shared physical boundaries, habitat corridors and environmental interests and the opportunity exists to create a UK wide network of habitats connected by shared corporate policy and commitments that would benefit the organisations individual and collective business interests as well as the UK biodiversity network.

8A.4 The opportunity for Intrinsic Earth

There are many shared environmental interests that business has the opportunity to develop as it faces its customers intrinsic connection to the natural world and the Intrinsic Earth Project (Intrinsic Earth, 2018) is in a unique position to facilitate the recognising and prioritisation of environmental and sustainability benefits.

- Intrinsic Earth can input and support business and organisations at different levels including:
- Start-up — niche business support, help through volatile early start-up and identify value of niche skills to communities and other organisations
- Mid sector — facilitating collaborative working, support through shift from technical to business focus. Similar to and/or including a Non-executive Director Network
- Corporate level: identifying environmental values of customers and staff and facing organisational change to strengthen business position through development of strategy and supporting policy.
- Linking common environmental interests and ecological resources of organisations and business' to create large-scale networks of environmental resources linked by a shared management commitment.

8B — Social Entrepreneurship in the Agrifood Sector: Smallholder Farmer Co-operatives

Sanjay V. Lanka

Sheffield University Management School, Sheffield, S10 1FL, UK

s.lanka@sheffield.ac.uk

8B.1 Introduction

I would like to thank Dr. G. V. Ramanjaneyulu (Ramoo) for taking the time to meet with me as a part of my research project on sustainable livelihoods in the Agrifood sector on a number of occasions to explain the CSA business model as well as for providing me access to the villages where the CSA works with the farmers to learn first-hand about how the business model works. I would also like to thank Ramoo for travelling to the University of Sheffield in May, 2018 to make a presentation at a workshop focused on the challenges facing smallholder farmers in India.

This contribution will focus on providing a narrative of the business model developed by the Centre for Sustainable Agriculture (CSA) in India, based on the principles of social entrepreneurship. The CSA has developed this business model by bringing together over 50,000 smallholder farmers to create farmers co-operatives, and a federation of these co-operatives in the form of a producer company that has created a chain of retail stores to sell the farmers produce directly to the end consumer. The CSA was developed in response to the challenges currently being faced by smallholder farmers in India in being able to achieve sustainable livelihoods. The inability of smallholder farmers to achieve and maintain sustainable livelihoods is due in part to what is known as the agrarian crisis.

8B.2   Background of the agrarian crisis in India

The Indian economy is structured such that, about 52% of the country's workforce remains dependent on the agricultural sector (Vaidyanathan, 2010). In India today, about 85% of the farmers are smallholders, and are defined as those operating less than two acres of land, with about 66% having less than one acre each (Sainath, 2011; Katakam, 2013). There are significant macro-economic factors impacting the smallholder farmers in India which need to be made clear to have an effective understanding of the context in which the CSA was started. Principal among these factors is the broad based agrarian crisis which has impacted many Indian farmers (Deshpande and Arora, 2010; Hebbar, 2010; Krishnaraj, 2006; Revathi and Galab, 2010; Sainath, 2011; 2013; Sidhu, 2010).

To understand the agrarian crisis in India, it is important to understand the *green revolution* which is a significant causal factor. Post-independence, the government of India followed Keynesian economic practices that involved investment in the public sector (Ahmed, 2011), which resulted in the institutionalisation of the use of chemical inputs which became the basis for the *green revolution*. As noted by Vaidyanathan (2010, p. 85):

> The past 50 years have seen unprecedented and far reaching changes in Indian Agriculture...per ha yields have shown a sustained and accelerated growth. A major contributory factor for this transformation is the rapid spread and increased usage of fertilizers. Total fertilizer consumption (in terms of nutrients) rose from a mere 60,000 tonnes in 1951–52, to 2.1 million tonnes in 1970–71. In the mid-1990s, it was around 16.5 million tonnes.

While the introduction of the *green revolution* has increased yields through the extensive use of inputs, it has also led to an increase in income inequalities among the rural sector and hence a further stratification among the social classes (McMichael and Raynolds, 1994). The *green revolution* has led farmers to be trained in so called modern farming — using pesticides, herbicides and genetically modified (GM) seeds that increase the input costs (Shiva, 1991).

The Indian government which had subsidised the use of fertilisers for over thirty years under its *green revolution* policy began dismantling the subsidy in 2010 and removed the regulation placed on the price of fertiliser (Anand and Chang, 2010). This was at a time when due to the ineffectiveness of fertilisers (Revathi and Galab, 2010; Reddy, 2010), over 15 million Indian farmers had already lost their lands due to indebtedness caused by the purchase of expensive inputs during the period 1991–2001 (Sengupta *et al.*, 2007). The reduction of agricultural subsidies leading to the high cost of agricultural inputs has placed a majority of Indian farmers in a situation wherein their costs of doing agriculture are greater than their income leading to the large scale displacement of people from rural India to the cities (Sainath, 2009; 2010; Sengupta *et al.*, 2007).

The Indian agrarian economy consists of a high percentage of its agricultural production being dependent on rain, a high density of agricultural workers consisting of smallholder farmers as well as agricultural labourers, and the structure of the land market is not set up in the interest of the farmer (Deshpande and Arora, 2010). The Indian agrarian crisis has been caused by an unfinished agenda in land reform, lack of water, technology fatigue; access to, and timeliness of institutional credit, as well as opportunities for assured and remunerative marketing (Deshpande and Arora, 2010). India's connection to the world market for imports especially those related to wheat and edible oil along with the increase in speculation through futures trading is driving food prices up (Shiva, 2008). This is leading to a high rate of inflation in the overall economy and impacting the real value of the subsidy provided by the Indian government for food security (Chandrasekhar, 2013). In turn, this inflation has a negative impact on the livelihood of smallholder farmers since it increases their cost of purchasing food.

The crux of the problem causing the agrarian crisis is that most of the value that is generated within food related value chains is at the retail end of these chains and the least amount of value is generated at the producer end of the chains (Daviron and Ponte, 2005). The value chain is a tool to analyse if each step in the supply chain of an organisation is generating any value for the end customer based on the actions of the employees of the organisation as well as its processes (Porter, 1985). The modern food supply chain is set up so that the farmers take all the risk, while large agribusiness companies that control access to the consumer at the retail level are able to buy food cheaply from the farmers (Jack, 2007).

8B.3 The need for social entrepreneurship in the agrifood sector

Smallholder farmers are impacted not only by production economics in terms of the cost of production at the farm level, but also by market economics in terms of the share of value retained by the farmer in the price paid by the consumer. At the farm level, the inputs costs for farmers are significant. However, farmers working with the CSA have replaced the use of these chemical inputs with systems of cultivation based on the principals of agroecology. Agroecology is the application of ecological principles and concepts to the design and management of agriculture (Gliessman, 2007), which helps mitigate the adverse impact of the practices of the *green revolution* (Altieri, 1983, 1993).

The chance to move away from the use of chemicals, and especially the use of pesticides was a huge motivation for the farmers to take up agroecological approaches to farming. The farmers were encouraged to make the shift by the problems caused by the use of chemicals in terms of soil degradation and the cost savings to be accrued from not using the chemicals. The net effect of eliminating the use of chemicals is that the total cost of production for the farmers has come down by 25%. Despite the savings in cost from not having to use expensive inputs, the smallholder farmers working with the CSA have had to use their own labour and resources on the farm to make up for it which has not been currently valorised at this stage of the value chain for agricultural products.

However, other costs have gone up especially the cost of land rental which accounts for 30% of the total cost of production. Another increase in costs is with regards to labour which represent about 30–35% of the total cost of production. Thus, for the smallholder farmer today about 60–70% of the cost of production is related to the cost of land and labour. These costs are a significant incentive to farm for those farmers who own their own land since effectively this decreased their total cost of production by 30% which would have been the cost of leasing the land if they did not own it. A related consideration due to the increase in the cost of labour is that the farmers consider the alternative of working as daily wage labour instead of taking the risk of farming on their own land or on leased land.

On the market side, the prices for agricultural produce offered to the farmer today are extremely low, irrespective of whether the type of cultivation is using chemical inputs or organic methods of cultivation. For example, during the year 2016 the prices for pulses have fallen by 50% as compared to the prices in 2015, from about $1000 per quintal to about $500 per quintal due to the opening up of markets allowing for a significant import of food from outside India (Bureau, 2016). Despite India being the largest producer of pulses with around 25% of global production in 2012–2013, its consumption during this year was 27% of global production and it had to import around 14% of its pulses requirements consisting of about 4 million tonnes (Singh *et al.*, 2016).

CSA had invested a lot of effort converting farmers who were growing cotton to take up the cultivation of pulses. But the large scale import of pulses from outside India has adversely impacted the prices for pulses in the Indian market which has led to a negative impact on the livelihood of the farmers who had shifted from cotton cultivation to growing pulses. In response to the call from the government that there was a need to increase the production of pulses to reduce the need for imports since nearly $2–5 billion had been spent importing pulses from Austria, Canada and Myanmar during the 2015–2016 fiscal year, farmers in India had increased the production of pulses in the year 2016–2017 by 35% compared to the previous year amounting to a record of around 22 million tonnes (Jadhav and Bhardwaj, 2017).

Thus the changes taking place in the market are outside the control of the farmer since the actions of the government play a significant role, as in the case of the import of pulses which caused the 50% price drop of pulses, leading to an increased vulnerability of farmers, while also reducing their risk bearing ability and make the life of the farmer precarious. On the other hand, making a shift to organic and agroecological systems of cultivation despite only reducing the cost of production by 20%, does increase the bargaining power of the farmer in the market in terms of product differentiation. So at least in this way the shift to sustainable farming methods despite not having all the solutions does provide some significant opportunities to farmers as compared to conventional systems of cultivation based on the technologies of the *green revolution*.

Most of the farmers working with CSA have organic certification through one of the following certification systems — participatory guarantee system of certification or the ICS third party system of certification. CSA provides the farmers with support in terms of ongoing training programmes to help them achieve and maintain organic certification. Further, CSA also conducts and maintains a system of internal audits to ensure the compliance of the farmers with the standards of certification. The data that is generated as a part of this process is maintained by the co-operatives of farmers which have been created and are the organisational structure that is the basis of the social movement initiated by CSA to improve the resilience of the farmers against the agrarian crisis. The entire cost of certification is borne by the co-operatives so that this is not a cost which has to be incurred by the farmers.

CSA has formed 30 co-operatives of farmers across the states of Telangana and Andhra Pradesh and all of these co-operatives together have formed a federation. This federation is set up as a producer company with the individual co-operatives as its members rather than the individual farmers. The producer company picks up all the costs including those for knowledge support, in terms of training that needs to be provided to the farmers as well as all the costs of data management and certification. The producer company also covers all the transactional costs including the costs related to procurement, starting at the farm level as well as all the costs related to licensing and taxation.

As previously noted one of the main challenges facing smallholder farmers is the fact that most of the value is created at the retail end of the food value chain. The solution to this problem has been developed by the CSA and this is to get the producer company involved at the retail end of the value chain. There are costs involved in setting up this retail infrastructure and the CSA producer company has to allocate 25% of the price at the retail level towards covering the retail and marketing costs. The remaining 75% of the price paid by the consumer at the retail level goes to the producer company. From this 75% of the retail price the producer company is able to cover the costs mentioned earlier related to knowledge support, data management, procurement, licensing and taxation and this comes to about 20–25% of the retail price. Thus the remaining 50% goes to the farmer which is a significant improvement as compared to the traditional value chains for agricultural produce where the farmer gets no more than about 20–23% of the price paid by the consumer. The innovation developed by the CSA in terms of creating this structure of farmers co-operatives and a producer federation that creates a retail entity has thus enabled the farmers to increase their share in the valorisation process from 20 to 23% of the total value generated in the conventional value chain to about 50% in the CSA created integrated producer owned value chain.

The basis for the ability of the CSA producer company to be able to provide this greater value to the farmers is due to their strategy of increasing the market price. The ability of the producer company to charge a premium price at the retail level is built on the bargaining power achieved from the product being organic certified as well as the traceability of the product along the value chain from the farm level to the retail store. This has enabled the producer company to address the conventional dip in the producer prices and increase the share in the value chain going to the farmers from 20% to 50% which is a significant improvement. This increase in the share of value achieved at the level of the smallholder farmer along with the reduction in the cost of cultivation due to the elimination of the costs associated with the use of chemicals has provided a significant economic benefit to the farmers to enable them to improve their livelihoods.

It is important to therefore understand the various elements in the innovation developed by CSA to promote social entrepreneurship in the Agrifood sector in India. The basic building block is the creation of co-operatives and the foundation of this is built on the farm school approach of the CSA to get farmers to move towards organic methods of cultivation.

A farmer field school consists of 10–15 farmers coming together and being trained every week in the field in terms of managing and understanding their farming over a period of four seasons. The four seasons are spread over a period of two years since in the Indian context each year has two seasons of cultivation namely the Rabi and Kharif seasons. Later as they complete the training in the farmer field school, these farmers become members of a co-operative and maintain the data related to their farming practices.

By continuing in the group they provide a form of collateral guarantee since if any one member of the group fails to follow the organic farming practices this leads to the whole group losing its certification. Thus consistent with the structure and ethos of self-help-groups (SHGs) developed in the field of micro-finance, these farmer groups are self-managed (Nair, 2005; Swain and Wallentin, 2009). A number of such farmer groups come together to form a co-operative. Each of these farmers SHGs are only registered as a group in the organic certification process as a part of the participatory scheme for organic certification set up by the government of India. Other than their recognition for certification purposes, this group of farmers is not a legal entity. The co-operative that is formed by a grouping of such SHGs of farmers is a legal entity and it is at this level of a co-operative that the farmers are able to hold membership in the producer company since the producer company consists of a grouping of a number of such farmer co-operatives.

Another innovation that is being developed by the CSA since 2016 is a logistics hub which is set up and shared between 4 and 5 co-operatives as a shared resource with the goal of providing them with infrastructure facilities. This hub is an amalgamation of the infrastructure needs of the farmers consisting of a warehouse, processing units, and packing facilities which essentially brings together all the logistics needs at one place. This hub is shared by these co-operatives and is owned by the producer company which is able to cover the costs related to setting up the hub as well as providing the resources including the management team needed to run the hub. This essentially provides the services that would be provided by a logistics and supply chain management hub where the farmers can store, process, pack and supply.

The challenges faced by smallholder farmers in India will require out of the box thinking. As evidenced in this chapter, the business model developed by CSA has been well designed keeping in mind all of the challenges faced by smallholder farmers and has developed innovative solutions for each of them. Of course, this does not mean that the problems facing smallholder farmers have been solved. Currently the CSA is directly involved with only about 50,000 smallholder farmers and there are over 250 million smallholder farmer households in India and many more around the world. Further, the unique set of challenges faced by women smallholder farmers have not been dealt with in this chapter and will require additional consideration. Thus, there is a need to scale up the CSA solution not only in the Indian context but at a global level. This presents an opportunity for likeminded people and organisations to replicate and customise the CSA model both within the Indian context as well as at a global level keeping in mind local needs and cultural context. In the long term, the problems faced by smallholder farmers will require grassroots bottom up initiatives using a social entrepreneurship model similar to the one developed by the CSA that incorporates smallholder farmers from the start and empowers them to fulfil their potential. The solutions for mitigating the challenges faced by smallholder farmers which have been outlined in this chapter have the potential to make a significant impact. However, in order for the momentum to continue, governments and other large scale funding bodies need to be made aware of the value and significance of these types of initiatives. Future work designed to engage with the CSA model and others like it should endeavour to involve a broad range of stakeholders, with priority given to issues around gender equity and social justice while seeking the participation of both governmental and non-governmental organisations.

Chapter 9

MOBILISING CITIZENS

9A — Dynamic Citizenship: Participatory Democracy on Local Level

Per-Eric Ullberg-Ornell

Formerly Municipality of Lidköping, Sweden

ullberg.ornell@gmail.com

9A.1 Introduction

This paper strives to describe a number of projects in Sweden and Bosnia–Herzegovina where the concept of dynamic citizenship has served as model for the democratic development. The reinforcement of the democratic process has been the overall perspective of the projects, where the relationship between the citizen and the decision makers is essential.

> In the democratic system, the citizen has the right to participate in every aspect of the political process. In fact, democratic development is a matter of concern and everyone's affair. "Users who are more involved in shaping the service they receive should be expected to become more active and responsible in helping to deliver the service" (Bentley *et al.*, 2005, p. 29).

The degree of political activity of the citizen can best be described by measuring the degree of political involvement in order of influence and the political process and the political decisions. Participatory democracy is about respect and confidence between citizens and the political system. It is important to create a mutual trust between citizens and decision makers and to be tolerant of different ideas that emerge in the political process. (Ministry of Justice Stockholm 2000, p. 1).

Dynamic citizenship consists of three "democratic qualities" — the degree of participation, the degree of influence and the degree of involvement. The meaning of participation is about a communication between the participants regarding common democratic values. Influence, on the other hand, is about real opportunities for the citizens to affect the political decisions. Involvement is about the possibilities for every citizen to act in different political contexts.

The three "democratic qualities" can only been reached if the terms of openness and insight, political equality and meaningful participation has been clearly defined as a political ambition. These three variables constitutes key concepts in order to the dynamic citizenship to be fulfilled (Ministry of Justice Stockholm, 2000, p. 1).

The definition of openness and insight can best be described as the development of increased accessibility to information and the ability to participate in the content of the policy decisions. Political equality refers to everyone's right to participate in social development regardless of ethnic or cultural background, social or economic status, sex or age. Meaningful participation is meant to give the individual citizen real opportunities for increased influence. If both participation and influence is perceived as meaningful, the sense of participation in various decision-making interest will be increased.

The dynamic citizenship can also be described in Figure 9A.1.

The process towards a higher understanding regarding democratic values and the importance of sustainable development is a complex history. According to a model presented by Brattberg (1995) shows the different steps of the process of change, which could be applied on the work with development of local democracy.

The first step is *information*. A person or organisation who is given information, *gives* something to the receiver. The receiver of the

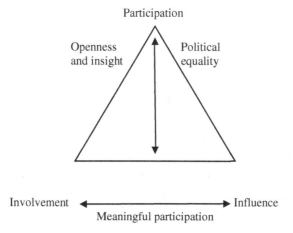

Figure 9A.1. **Dynamic Citizenship Model (*Source*: Ministry of Justice Stockholm, 2000)**

information has to *take in* the information in order to turn it into *knowledge,* which is the second step in the model. Already here is an obstacle, i.e. the attitude to the information of the receiver. People act irrationally and in that way they are unpredictable in how they are receiving a communicated message. In order to succeed in that aspect, the information should be seen as *useful* and that it is *supporting* initiative initiated by the receiver. The information has to be balanced and should not be interpreted by the receivers' demands, which inhibit the motivation to change.

In this case the concept *attitude,* the third step in the model and in the process of changing, means a lasting attitude to something based on experience of democracy. It is built up experiences and will be reflected in the fact that you are positive or negative. Everybody has ideas about how the society should be and who are running the society. That is shown by the complexity of opinions and actions in the projects above.

A final step in the process of changing is *behavior.* A result when information leads to increased knowledge and attitudes. It means that the citizens and the decision maker goes to action with equal social norms and a common view on democratic values.

9A.2 Case studies

The following chapter describes the four case studies regarding local democratic development and where the model of dynamic citizenship has served as point of departure. Point of departure for all four cases has been the above described model, with the overall aim to identify whether the model can be used in projects with different democratic approaches.

The first case is from the municipality of Lidköping, Sweden and was conducted during the period of 1998–2001. The two following cases where conducted in Sarajevo City (Bosnia and Herzegovina). During the period 2001–2014. In the fourth case was conducted in a biosphere reserve in Sweden during the period of 2011–2014.

9A.2.1 *Case study I — This is our future! Vision process Lidköping*

The project "This is our future", was created in the municipality of Lidköping (Sweden) during the period of 1998–2001 (Lidköping Kommun, 2001, pp. 9–13). The process was initiated by the local politicians. And as based on concrete actions in order to visualise the concept of democratic development, focusing on social, economic and environmental dimensions of sustainability. The overall aim was to let all citizens participate in the process in order to shape the future for the municipality in a participatory process. It meant that individual citizens, different NGOs, interest — groups, private businesses and people working in the public administration, took part in the process.

The citizens, their views and reflections, served as baseline as well as point of departure for the following process. As a result of this approach, and with the slogan "the center is the citizens", the project produced a large number of opinions regarding sustainable democratic development on a local level.

The working method in the project was divided into three phases: structure (1), process (2) and results (3). The structure phase contained the following elements; internal and external anchoring of the project, relevant political decisions, project organisation, collecting facts, information and dissemination of project objectives and benchmarking.

The process part of the project included concrete activities towards educational organisations, local business and industries, civic organisations and so forth. This part of the process also included different organised study — circles, hearings, seminars and workshops, written articles in the local media and continuous and relevant information on the web page of the municipality.

The third step contained the results, as in visions and strategies. More than 250 visions and over 500 strategies were formulated by the citizens. They described, in a very concrete sense a sustainable society, defined by the citizens themselves. All opinions were documented, and later on transformed into a vision and strategies (A *vision*, in this context, is a text describing a direction towards the three dimensions of sustainable development. A strategy translate the visions into a direction of concrete wills.). The content in the vision as well as in the strategies, was formulated by the participants themselves.

The document *This is our future!* Was adopted by the local politicians in the municipality in 2001/2002 and became an important steering document for further plans and policies for the municipality. The vision document was also published as a book and distributed to interested inhabitants living in the municipality of Lidköping.

9A.2.2 *Case study II — Dream city Lidköping — Sarajevo*

Bosnia and Herzegovina was, in the middle of 1990s a country in chaos. Post-war implications such as bad economy, bad living conditions for the citizens and high unemployment rates, especially among young people, showed the need for relevant reforms in order to change the process into a positive development instead.

"The Dream City pilot-project" was initiated in 2001 by the municipality of Lidköping and the municipality of Centar in Sarajevo City. The project was also supported by SIDA. The overall aim of the project was to inspire a positive change, simply by increasing the knowledge of citizens and decision makers regarding the importance of common democratic values. The aim was therefore to strengthen the citizens understanding and engagement around the necessity of a common commitment in order to reach a democratic society (Filipovic and Ullberg-Ornell, 2003).

By using the model of dynamic citizenship, connecting it to the three dimensions on sustainable development, the Dream City project tried to change the development in Sarajevo into a positive development. The purpose was to identify possibilities as well as obstacles in order to create an environment defined by equal societal responsibility. In that sense the project tried to implement new perspectives experiences, knowledge and ideas around the meaning of democracy.

The results of the pilot-project showed that the citizens' participation as well as attitudes towards democratic values was very low. The results could be explained as an outcome of by the very long period of time without democratic governance in the country. The knowledge and experiences of how a democratic system can function in practice had been subordinated due to an authoritarian acidification of the county throughout the past centuries. The awareness of this historical based situation was one of the biggest challenges during the project period. In order to overcome the situation, the people involved in the project worked closely with different international organisations in the area, already working on different types of democratic implementation in Sarajevo.

Another result from the project was the process of change. In this case, the main difficulty and challenge was to transform from one "system" to another. It does not happen automatically. Implementation of democratic values by individuals as well as organisations need consistent work, using different tools and combining them into theory and practice.

One way to handle this type of situation was to use *simple messages* in combination with *concrete examples related to practice*. Even if the citizens were not always ready to make immediate change directly, the process continued in order to make time for change to occur and to be sustainable. In other words, focus where put on a combination between *identity* and *engagement* in the process. It meant that the citizens and involved institutions were given relevant and sustainable tools and information in order to be able to work with democratic values and implement them into a political system based on equal terms for all (Filipovic and Ullberg-Ornell, 2003, pp. 8–18).

9A.2.3 *Case study III — Enterprise democracy*

The concept of "enterprise democracy" is taken from a report made by DEMOS, a think tank in the UK (Bentley *et al.*, 2005). On behalf of the Swedish government, they published in 2005 a report describing a future scenario where Sweden can compete and be successful in a globalised economy. To make that real, DEMOS are suggesting a stronger co-operation between the business sector and the civic society. By using the creativity from both parties, the report indicates a model for sustainable growth where participation and shared ownership is in focus. The platform for this development is, and as mentioned above, the so-called Enterprise democracy (Bentley *et al.*, 2005, pp. 28–33).

Based upon the report made by DEMOS, a second project was launched in Sarajevo in 2011–2014, initiated by the municipality of Lidköping and the municipality of Novi Grad, Sarajevo City. The project focused on the concept of enterprise democracy, and where the model of dynamic citizenship was implemented. The project focused mainly on the correlation between democratic values and a societal structure based on sustainable growth.

The project assumed three perspectives in the context of enterprise democracy. The first perspective was the importance and meaning of social entrepreneurship. Social entrepreneurship is a process where the citizens have the possibility to start new businesses or institutions with main focus of solving socioecological problems in society. The idea behind social entrepreneurship is to establish new companies in areas where the public sector have not been able to solve (or handle) the problems. For example, companies within the context of social entrepreneurship, could be companies working with poverty, human rights or environmental issues.

The second perspective was social innovation. Social innovation can, in brief, be described as a democratic approach towards unused resources in the society. It means innovative services and products addressing new ways to challenge social or ecological structures.

The third perspective contained social capital as key to a successful result. The interpretation of the term social capital has a lot of different meanings according to academic science and theories. In this case following definition on social capital has been used and added to the model of dynamic citizenship. In this context it means the following:

> "Social capital is expressed thought trust crated in relations among people. Social capital exists among people and evolves over time. Social capital is a resource existing in social structures and created by human action." (Bergstrand and Ullberg-Ornell, 2009, pp. 28–33).

Different perspectives on social capital was raised in the project. The aim was to strengthen the relationship and dialogue between people from different contexts. The purpose was to overcome conflict and mistrust which, during the period for the project, still characterised the climate in the local society.

Taking these perspectives as point of departure, the project managed to engage people in new discussions, and also made it possible to approach social entrepreneurship as an incentive for renewal of the structure of society. The outcome of the project was that a good relationship across different boundaries (such as gender, for example) is necessary for a successful result. In that aspect, the project became, by the participated parts, a learning process emanating new knowledge and ideas how society can be formed, ruled and progress (Ullberg-Ornell, 2014).

9A.2.4 *Case study IV — Biosphere entrepreneurship*

The fourth case study, is the project "Biosphere Reserve Vänern Archipelago and Mount Kinnekulle in Sweden". This study is a good example of specific version of sustainable entrepreneurship, namely called biosphere entrepreneurship. It also describes an innovation system based upon on democratic values. In short, the model be summarised as a sustainable participatory economic development.

The Biosphere Reserve Vänern Archipelago and Mount Kinnekulle project had an ambition of creating a model area for sustainable growth. The meaning of biosphere is in this context, the possibility to connect core values of sustainable growth with ecosystem services to improve the well-being for every citizen. Natural ecosystems in a biosphere area is a unique resource, used by new companies with a high degree of social benefit, called biosphere entrepreneurship.

Biosphere entrepreneurship is characterised by core values such as improvement of human welfare in the context of an ecosystem. The companies also have a high degree of independency and the financial profits will be reinvested in the area where they are situated. In other words, a biosphere entrepreneur has a democratic approach to making money (Bergstrand, Björk and Molnar, 2011).

Using the meaning of dynamic citizenship as point of departure, the project created an innovation system for biosphere entrepreneurship. An innovation system, in the context of a biosphere reserve, is a supporting system for ecosystem services. The innovation system consists of four different interconnected perspectives. They are universities, financial institutions, public institutions and organisations and the civic society (people living in the area).

There are two things that make this type of innovation system unique. One is that the citizens have an important role as co-creators in the process of new companies. The second one is that the companies are scalable on local as well as on global level. It means that the biosphere innovation system offers possibilities and forums for actors willing to cooperate. The awareness of the democratic approach towards sustainable growth is one of the core perspectives bringing people together within this system (MacTaggart and Ullberg-Ornell, 2012).

9A.3 Conclusions and reflections

The value and implementation of the model of the dynamic citizenship is the common denominator of the four above presented cases. The cases describe how this model can be used in different processes of change, with the development of democracy as the main goal.

The projects themselves and the model of dynamic citizenship clearly showed the importance of trust between citizens and decision makers in order to create democratic development.

The projects showed that increased trust can lead to dynamic citizenship at all levels.

When starting local democracy development process, it's necessary to focus on an open mind relationship between citizens and decision makers. It is important to prevent mistrust by asking questions like *what to do* and *why it should be done*. These questions, regardless of the context (Sarajevo or Lidköping), will make it possible to start a process showing the potential participants, a type of "democratic design" perspective, where the importance of collaboration is essential and in focus.

It is important to be aware of that "your own intrinsic interests", as a citizen or a decision maker, is not enough to achieve a sustainable democracy development on local level. Sharing and listing to experiences made by others before taking action is essential.

The opportunities to achieve a good result increases if the model for dynamic citizenship is used. The overall results from the projects, described in this article, shows that common interest and commitment is closely associated with common democratic values.

Most of the projects mentioned in this article, shows the importance of social capital as an underlying structure in a democratic process. To invest in human capital, by working with the meaning of social capital, requires a transparent approach between the participating parts. It means that the structure must contain a high degree of flexibility, tolerance and social trust to obtain an equal ownership (Brattberg, 1995).

The awareness of that social capital is about relationship/network, trust and acting together have a great impact on the results in a democratic process. The experience from the projects have showing that this leads to growth and democratic stability. (Bergstrand and Ullberg-Ornell, 2009).

The experiences generated by the different projects in this chapter and in connection to the methods and model of dynamic citizenship can be summarised as follows:

- The importance of concrete activities where knowledge as well as experience are valued equally.
- In order to create a real democratic process, it is important to let the involved actors describe themselves what constitutes democratic values. It should also be done in the beginning of a democratic process.

- All actors involved need to take responsibility for the progress of the democratic process.
- Shared ownership added to identity and engagement.
- Communication between different actors are a critical factor to create common commitment regarding the essential democratic values.

9B — Engaging Imaginaries: The Role of Artistic Collectives for Transdisciplinary Sustainability

Andressa Schröder

Justus Liebig University Giessen, 35390, Giessen, Germany

andressa.schroeder@gcsc.uni-giessen.de

9B.1 Introduction

The focus on transdisciplinarity has grown within the academic community in the past two decades. There have been increasing publications about the integration of transdisciplinary forms of knowledge in relation to different topics and fields of study (including sustainability), the development of a transdisciplinary methodology, and the definitions of transdisciplinarity itself. There are two main approaches to transdisciplinarity which are usually referred to in the multiplicity of papers referring to the topic. One is the "Nicolescuian School" and the other one if the "Zurich School". Both approaches deal with complex issues of the contemporary world and challenge imposed disciplinary boundaries to tackle such issues. However, they differ fundamentally in their philosophical background and understandings of methodological procedures and aims for transdisciplinary research (Bernstein, 2015). In this paper, I do not intend to deepen the discussion about the differences and/or similarities between these approaches and I follow mostly Nicolescu's thoughts about transdisciplinary methodology.

Even though there is not a unique and unanimously accepted definition for it and sometimes transdisciplinarity is considered as an ambiguous or non-reliable approach to "scientific" matters (Lang *et al.*, 2012), it brings a necessary dialogue and democratisation of knowledge into the realms of scientific debate and an openness for how to frame different social and scientific issues, as well as the possibilities for finding solutions for them.

In a general sense, transdisciplinarity refers to transgressing the borders of disciplinary (scientific/academic) knowledge and incorporating different forms of knowledge as valid and important elements for research. This happens not only in the sense of interdisciplinary collaborations, but also in the sense of opening research for non-academic forms of knowledge. Not only incorporating the works of different practitioners as valid forms of knowledge, but also engaging with multiple forms of local knowledge and practice — the "lay knowledge".

Within sustainability studies and debates, transdisciplinarity has been explored as a fundamental feature for meaningful actions, projects and decisions (Schäfer *et al.*, 2010; Jahn, 2008). It is an important approach to combine theory and practice, or research and praxis (Evans, 2015). Even more importantly, it moves sustainability research beyond scientific matters to engage with ethical ones. As highlighted by the scholar Tina Lynn Evans, who focuses on sustainability education:

> Transdisciplinary faculty collaborators must be able to live and work in a space of ambiguity, recognizing that sustainability is a process, not an end. Humility, a willingness to learn from others (including those outside of academe), and a personal identity that does not derive heavily from one's authority as an expert are typical characteristics and orientations of successful transdisciplinary educators. When engaging with sustainability praxis, we must be willing to admit that, while what we know is valuable, what we do not know is immense. We must recognize that all windows on knowledge and the world offer only partial views" (2015)

Evans highlights the role of transdisciplinarity for sustainability education, but this is not the case only for education. In any approach to sustainability it is necessary to have respect for multiple worldviews and forms of knowledge. It is important to realise that sustainability is also defined by contextual (ecological and cultural) circumstances and that there is not a singular formula for right or wrong approaches to it.

Another important characteristic of transdisciplinarity is that it is not necessarily solution-oriented. It may depart from finding and discussing specific problems, but the process involved in enabling dialogue and multiple perspectives to arise is more important than finding technical solutions for problems (Evans, 2015). This is not to say that solutions are unimportant for transdisciplinary approaches to sustainability. On the contrary, it is realised that solutions are dependent on processes of communication and ethical decisions that respect multiple perspectives and understandings of such problems and go beyond imposed ready-made models.

Arts-based research can be closely associated to such premises of transdisciplinarity. It is process-oriented and often deals with flexible frameworks that are not limited by the methodological steps of a single discipline. It engages creativity, sensitivity, perception and experience as valid and crucial forms of knowledge. It also often, mainly in contemporary practices of environmental artistic research, involves multiple processes of collaboration — from the conceptualisation, through the realisation of a project, to the expression of different outcomes for it.

Similarly, environmental aesthetics revolves to multi-sensorial experiences of the environment, which can be very subjective but can also form collective perceptions and imaginaries. Environmental aesthetics extends way beyond the superficial qualities of landscape appreciation. It involves multiple processes of perception, imagination and engagement, and it reaches to the formation of environmental awareness and the realisation that human beings are interdependent elements of complete ecosystems.

Basarab Nicolescu proposed three axioms to develop a transdisciplinary methodology:

I. **The ontological axiom:** There are, in Nature and in our knowledge of Nature, different levels of reality and, correspondingly, different levels of perception.

II. **The logical axiom:** The passage from one level of reality to another is insured by the logic of the included middle.

III. **The complexity axiom:** The structure of the totality of levels of Reality or perception is a complex structure: every level is what it is because all the levels exist at the same time (2006: 9).

By the logic of the included middle, Nicolescu emphasised that no form of knowledge is self-enclosed (*ibid.*: 18). Thus, transdisciplinarity refers also to the knowledge that lies in the gap between disciplines. It is the space between, across and beyond disciplines (*ibid.*: 5) — the void of non-disciplines. In some sense, arts-based research can be understood as fulfilling this void similarly to transdisciplinarity. It engages with different topics and explores them combining methods from various disciplines, but it does not have to necessarily give any of these disciplines back a concrete product or solution. It enhances various forms of aesthetic experiences in relation to the topics of such disciplines and it stimulates the expression of distinct perceptions connected to them. It also stimulates revisiting old perceptions that have already been expressed and reconsidering them under renewed experiences and circumstances.

There are also different interpretations for what arts-based research means or what kind of methodology it implies. Some interpretations and uses of arts-based research may be found in the examples of Barone and Eisner (1997) and Irwin and De Cosson (2004) (in relation to art-education); in Pickering (2008) and Knowles and Cole (2008) (in relation to social and cultural sciences), or in McNiff (2013) (art therapy). The one that is under focus here is what I have defined elsewhere as *research through art*:

> In research *through* art, the artistic-thinking-process is understood as a methodological procedure in itself and the different stages of the creative process can be identified as distinguished methods for the development of the research (Irwin & De Cosson, 2004; Sullivan, 2005). [...] the creative thinking process, although it may be chaotic and subjective, is also systematic (Sullivan, 2005) and can be incorporated in the process of research related to various fields." (Schröder, forthcoming).

In order to investigate the potentials of arts-based research and environmental aesthetics for a transdisciplinary sustainability in a more pragmatic form, the article follows with the exploration of the example of the work of the Brazilian artistic collective *Coletivo Líquida Ação* and more specifically their project *Foz Afora* developed in 2017.

9B.2 *Coletivo Líquida Ação*

The *Coletivo Líquida Ação* is a fluid group formed by various artists with different backgrounds who research artistically the topic of water in relation to space and bodily memory. The artists experiment with different artistic media (visual and plastic arts, video and sound, poetic language and scenic techniques) culminating their work in artistic interventions and performances that usually occur on the streets, squares and other public urban spaces and that are open for the direct interaction with the public (Figure 9B.1).

The collective was formed in 2006 and ever since the dynamic in the group has been fluid. According to Eloisa Brantes Mendes, coordinator of the collective, the roles performed within the collective as well as the choice of participating or not in the artistic actions and interventions depart from the artists themselves and the group's dynamic is established with no hierarchical impositions. (Eloisa Brantes, granted me a very

Figure 9B.1. *Ouro Líquido-Mitologias Urbanas* (Liquid Gold-Urban Mithologies). (*Source*: Coletivo Líquida Ação, 2011).

enlightening Skype-interview in January 2018. We discussed several aspects of the formation of the artistic collective, the projects *Volume Morto* and *Foz Afora*, and hers and the collective's experiences of the situation at the *Doce River*'s estuary. This is unfortunately not a published document, but the references are cited as Mendes (2018).

This fluidity taken as a core element of the collective relates closely to the liquidity of the main theme explored by the group — water. Water is a vital element for life, in constant movement, fluid, liquid and malleable. Furthermore, our bodies are formed by 70% of water, which characterises the micro-dimension of this vital relationship. By combining the fluidity of experiential and bodily memory with this fundamental element for life, the group explores subjective experiences and goes beyond them to the emergence of collective imaginaries formed around this substance:

> In the intervention-performances the uses of water intensify the temporality of the body-space relation in two aspects: the immediate reactions of the body in contact with the water and the access to the collective memory about water in the cities. (...) It is a transpersonal corporeality, that is, a body-performer that transits through different layers of subjectivities produced in the time-space that constitute the performance. (Mendes *et al.,* 2017: 127–128)

The multiple temporalities and transpersonal corporeality explored by the collective reflect a multiplicity of perceptions and come in this sense close to the understanding of multiple realities emphasised by Nicolescu in his axioms for a transdisciplinary methodology. "The concept of "liquid action" echoes at the intersection of these distinct temporalities; that is, a malleable action which fluxes, forms and paths depend on the physical and symbolic supports of space, the bodies involved [between, across and beyond this space] and the interactions of various social and cultural values that operate in performances." (Mendes *et al.,* 2017, p. 128)

The name of the group also implies a homophone in Portuguese holding its more direct meanings in *líquida ação* (liquid action) and alluding to the meanings of *liquidação* (liquidation/sale off) (Figure 9B.2).

In this sense, the different values attached to the substance of water can be investigated and challenged. The overexploitation and commodification of water affect the meaning that it acquires in the collective

> **Liquid action** - 1. Action that flows or runs, taking malleable forms in the time/space of its context; 2. An action whose value is not subject to tax reductions or rates; 3. Action in aliquid state; 4. Action in which the intensity of human and molecular interactions are unpredictable; 5.Vital action of water.
>
> **Sale off (Liquidation)** - 1. Activity required to renew stocks; 2. Liquidity of products by reducing their financial value; 3. Great celebration of consumption, which collective excitement generates strong sensations of pleasure, anxiety and euphoria; 4. Characteristic phenomenon of the consumer society that offers immediate profit to the attentive consumer.

Figure 9B.2.   *Líquida Ação/Liquidação* (*Source*: Coletivo Liquida Acao, 2018)

imaginary of different communities and social groups. Its vital significance is transformed into a product and its value is governed according to the economic dynamics of the market.

The collective investigated these subversions of the value of water in a project entitled *Volume Morto* (which later also developed into the project *Foz Afora*, analysed in the sequence of this chapter). The expression *volume morto* (which translates as dead volume) is used in Portuguese to refer to the technical reserve of water that lies below the intake pipes that are normally used to draw water from dams. This volume of water is considered "dead" because it should not be used due to the concentration of high levels of heavy metals contained in it. Besides, and more importantly, this "dead volume" should not be used because it holds a fundamental value for the maintenance of the local ecosystems.

In 2014 there was an intensive hydric crisis in big urban centres in the Southeast area of Brazil, mainly in the region of São Paulo and the use of the "dead volume" became more constant, which made the expression more popular as well.

"The "uselessness" of the water reserve is fundamental in the dynamics of the functioning of springs because it maintains the whole ecosystem of aquatic life. The consumption of water contained in the dead volume can cause irreversible damage to the environment. But during the water crisis an accumulation of factors — the despair of the population, the unpreparedness of public officials and managers in dealing with the "announced tragedy" by their own administrative inefficiency, and the media sensationalism in denouncing the waste of water and asking people to save on their domestic consumption — it became more important to use the water of the dead volume than [to preserve] its vital permanence to the ecosystem. The reversal of this utilitarian perspective on reserves was the starting point of the creation of the show [*Volume Morto*] based on several fragments of personal and collective archives as activators of our water-life reserves." (Mendes *et al.*, 2017: 128)

The more the hydric crisis worsened, the more it affected directly the activities of the collective. The collective was restricted from using water during their performances and the artists started exploring other formats and materials that would enable their investigation about this element to evolve: "In dealing with such a "restriction" in the collective action of transporting water throughout the city, the group (re)affirmed the nature of its research and updated its creative course in performances through the encounter with the flow of events. How to evoke the liquid, vital, imagetic and sensorial power of the element water in its absence?" (Chilinque in Mendes *et al.*, 2017, p. 130) They moved on to using sand and other elements that, in a close relation to their bodies, would activate their memories in relation to the substance of water: "At the time, the choice for sand was not a representation of the lack of water, but the power of water in its absence, modifying actions and the guidelines of the intervention proposal. Considering the change of weight, the sand modified our relations with the gravity and with the density of the relations between the body and the city" (*ibid.*).

Soon the project *Foz Afora* emerged as an extension of the *Volume Morto*. The collective wanted to investigate the estuary and basin areas of the *Doce River* in the Southeast of Brazil. The river's estuary was suffering the consequences of this hydric crisis and slowly and dramatically drying out. Though, amid this process of predation, another tragic event affected the basin of the *Doce River* and before going into more details about the project *Foz Afora*, I want to briefly contextualise the situation and tell the story of one of the biggest cases of environmental impact in relation to mining activity in Brazil and probably in the world (Carmo *et al.*, 2017; Neves *et al.*, 2016).

9B.2.1 *The death of a river*

On the 5 November 2015 an iron-mining dam run by *Samarco Mineração S.A.* (a joint-venture formed by the Anglo-Australian BHP Billiton and the Brazilian Vale S.A.) collapsed unexpectedly releasing around 50 million cubic-meters of metal-rich tailings waste (IBAMA, 2015).

The dam, *Barragem do Fundão*, is located at the municipality of *Mariana* in the state of Minas Gerais, Southeast of Brazil and the extreme volume of tailings immediately devastated the subdistrict of *Bento Rodrigues* leaving 19 people dead and hundreds of families displaced. The wave of mud flew around 77 km through the *Gualaxo do Norte* and *Carmo Rivers* affecting other villages in the area and finally reaching the *Doce River* after 48 hours (IBAMA, 2015). Around 41 municipalities along the *Doce River* basin in the states of Minas Gerais and Espírito Santo were directly affected by the mudflow. It is estimated that more than 1 million people in the areas along the river were direct or indirectly affected with the lack of access to clean water and other natural resources, lack of hydroelectric power generation and restrictions for fishing practices as well as crop production (Fernandes *et al.*, 2016).

Furthermore, it has been indicated that the tailings directly affected "135 identified semideciduous seasonal forest fragments, in a 298 hectares of vegetation suppression, located in the banks of the *Gualaxo* and *Carmo Rivers* and its tributaries. The tailings also directly hit 863.7 hectares of Permanent Preservation Ares associated to watercourses" (Carmo *et al.*, 2017, p. 147). The *Doce River* is part of one of the Atlantic Forest biomes, which has a level of biodiversity as rich as the Amazon Forest. Several endemic species of animals and plants where severely affected by this environmental catastrophe (IBAMA, 2015).

It took 16 days for this toxic mud to flow more than 600 km down the *Doce River* and reach the Atlantic Ocean on November 21 (IBAMA, 2015). Finally, "[d]ue to the action of northward ocean currents in the Atlantic, fine suspended sediments have spread through marine habitats of the Brazilian coast. Consequences at broader spatial scale, including international waters through the transboundary movement of suspended sediments, remain largely unknown" (Fernandes *et al.*, 2016, p. 36).

Besides all the biological and ecological devastation, this environmental accident/crime also caused a great loss of cultural heritage, including archeological sites, caves, places of historical and cultural interest and areas with economic interests related to tourism, and cultural practices and celebrations. Several riparian communities, including indigenous groups were physically, socially, culturally and spiritually affected (Carmo *et al.,* 2017; Neves *et al.,* 2016; IBAMA, 2015).

The company, *Samarco*, has been fined by the Brazilian Institute of the Environment and Renewable Natural Resources (*Instituto Brasileiro do Meio Ambiente e dos Recursos Naturais Renováveis — IBAMA*) and the Governments of the states of Minas Gerais and Espírito Santo several times and in March 2016, an agreement for a 15-year revitalisation plan estimated in 5.2 million dollars was established. In May 2016, the prosecutor's office issued a new process in which the damage was reevaluated in 43.4 million dollars (Dornelas *et al.,* 2016; Lopes and Wernek, 2017). However, the company has appealed and up to August 2017 only 1% of the amount had been paid. Besides that, several judicial motions against *Samarco* by people who were affected by the accident/crime have been temporarily suspended (Bedinelli, 2017).

9B.3 *Foz Afora*: Artistic residency at the *Doce River* estuary

The project *Foz Afora* was written by the *Coletivo Líquida Ação* for an open call for artistic residencies before the above mentioned environmental accident/crime happened. (The call was for the 2015–2016 version of Rumos Itaú Cultural, which is an annual funding program for Brazilian artistic and cultural projects financed by the non-for-profit Institute Itaú Cultural.) More as previously mentioned, at that point it was part of the project *Volume Morto* and it aimed at investigating the basin of the *Doce River* because it was already suffering the consequences of intensive and increasing predatory use of natural resources. The project was submitted by the collective to the selection process only a few days before the catastrophe took place in 2015.

After the artists were notified about the acceptance of their project for funding, they decided to restructure the project to incorporate the current situation of the river, its ecosystems and the lives of the people that were affected by the catastrophe into the project. This initial reconsideration of the project according to the current circumstances of the "object" of research was possible because of the flexibility of the Program *Rumos Itaú Cultural* and it is already a crucial step in the process of arts-based research. It demonstrates that such a project cannot be completely pre-defined or stagnated. It must embrace flexibility and even certain levels of instability.

Following the renewed project, in June 2017, nine artist-researchers went to *Vila de Regência Augusta* (further referred to in this article as *Regência*) to begin their residency at the *Doce River*'s estuary, where it encounters the Atlantic Ocean in the State of Espírito Santo. They travelled along the *Doce River* in the trainline Vitória-Minas, which connects the states of Minas Gerais and Espírito Santo. The artists report their experience of this 13-hour trip in an online diary: "From the window I see endless freight trains passing by. Each wagon carries two and a half tons of ores. Each train displaces 280 wagons. How many Vale's freight trains leave daily from Minas Gerais towards the port of Vitória? Small cities tied to the 52 stations seem to emerge from nowhere. Iron rails, railcars and mining dams cut through the silent nature. Earth drilling is a corporate rape." (Mendes *et al.*, 2017).

The artists spent then 22 days closely accompanying the lives of members of the communities of *Regência*, *Povoação* and *Areal* all based at the estuarial area of the river. Departing from the question "How can the immersion of artists-researchers in this traumatic reality participate and contribute to the re-elaboration of the collective imaginary?" (Foz Afora, 2017). The artists aimed at investigating how the relationships in these environments worked and consequently how such a tragedy affected their cultural practices and collective imaginary.

The artist-researchers, each following their own personal encounters, collected various narratives of the members of local communities, made videos and photos, collected oral and visual material for artistic installations and scenic experimentations that would be later developed. They also participated in local celebrations, learned about local practices like traditional fishing techniques and the production of fishing nets, and they collaborated with artistic initiatives of local groups. Through this immersive process the artists investigated the multiple levels of the symbolic violence that has for long accompanied the destructive use of natural resources along the *Doce River* and culminated in the tragic accident/crime in November 2015. "The landscape of the estuary of the Doce River is as

DA ÁRVORE NASCE A REDE?

FOTO: JERÔME SOUTY

Figure 9B.3. *Do Nets Grow on Trees?* (Souty, 2017)

material as imaginary, it is shaped by individual and shared memories, as well as by projections and representations. To the inhabitants of the region, the coexistence with the natural elements is combined with the emotional memories of the bygone river." (Souty, 2017, p. 49) (Figure 9B.3).

According to Eloisa Brantes, the lives in these villages have been fractioned due to the lack of access to clean water. The complex activities that were related to the river and that required collaborations and collective work have now been replaced by an undermined financial compensation (Mendes, 2018). Many families that "can prove" that they have been affected by the disaster receive a card that entitles them to get financial support from the company *Samarco*. Nonetheless, several families have had their judicial motions suspended (Bedinelli, 2017), which makes this process of compensation unclear and unfair. Furthermore, this financial compensation is not enough to fulfil the complete cultural and symbolic loss that the communities suffered with the devastation of the river. The water of the river was the point of connection among different people and families that shared the practice of fishing, surfing, and swimming. Its value goes beyond the economic subsistence that it could provide. The activities that surrounded the river cannot be reduced to the payment of a monthly salary. This concern had already been indicated in the first technical report about the accident/crime issued by IBAMA: "The physical separation of neighbours and groups from a community causes people to lose their traditional, cultural, religious, and local identities, bringing disruption to their intrinsic and intangible values, which are not healed by the distribution of kits, indemnities, or the rent of houses in other neighbourhoods" (IBAMA, 2015).

The fragmentation of the collaborative practices and collective identities of the communities holds also a political danger of segregating their interests and obfuscating their voices as communities that depended on this river on multiple levels. When their "collective life is fractioned" (Mendes, 2018) and their vital relation to the water is reduced to a form of financial compensation (in a process that is not transparent), there is a wide margin for disagreement and mistrust. Therefore, the work of the *Coletivo Líquida Ação* brings a fundamental connection through bodily memory and its relation to spatial perception. By engaging with the shared memories that were intrinsically connected to bodily practices, the local people as well as the visiting artists could experience and share, old, new and renewed perspectives of their own relation to the river. Re-evaluating and negotiating the significance that the river held for their lives, cultural practices, spirituality and survival. Consequently, strengthening their voice and belief that the river and surrounding ecosystems hold a value that goes beyond their economic subsistence.

Despite the intensive immersion of the *Coletivo Líquida Ação* in the village of Regência, the artists did not intend to engage in political militancy. Instead, they wanted to find more neutral forms of engaging with the local populations without any pre-established intentions. They aimed rather at listening to the local stories and voices instead of bringing them a ready-made ideological discourse (Mendes, 2018). This does not mean that arts-based research cannot become a powerful force to indicate problems and demand changes. This can occur and can generate great outcomes, but it is not the ultimate goal of arts-based research. In the case of the local interventions of the project *Foz Afora*, for example, the artist-researchers had varied encounters with members of the local populations and engaged in (or in some cases initiated) several kinds of small mobilisations through the investigation of the multiple layers of bodily memories entangled with environmental perception, as well as self-reflection and critical reflection about the current cultural and environmental situation that they are forced to live.

This form of research also differs from other forms of scientific research because it does not search for specific pre-established results. It focuses on the processional characteristics of the encounters with various situations, like the narratives and memories shared by different members of the communities, the collective imaginary that is unveiled through these narratives and, of course, the artists' personal experiences in this environment:

> As artists and researchers, we propose displacements and reflections about the experiences of the sensible in this context of environmental catastrophe. We entangle field research and artistic residency, develop articulations between environmental, social, and political engagement and artistic experimentations in different moments and places [...] Our role is not to search for technical solutions to depollute [the river], nor for forms of political or juridical organization to face the environmental disaster-crime. We use the notion of device to activate reflections and subjectivities from physical and symbolic displacements (Souty, 2017: 20).

This "methodology of displacement" is defined by the artists in four different stages of their research: the first one is the railway trip along the *Doce River*; the second is the artistic residency at the river's estuary; the third refers to the artistic experimentations developed with the collected material (like the production and edition of videos, images, sounds, as well as the participation in local cultural practices); and the fourth concerns the scenic experimentations and artistic installations presented at the theatre and gallery of the *Espaço Cultural Municipal Sérgio Porto* in Rio de Janeiro towards the end of the project (Figures 9B.4 and 9B.5).

In all stages of this "displacement methodology", the artists-researchers engaged in multiple levels of transdisciplinarity. In a first instance, each of the members of the collective had their own experiences, perceptions and expectations that had to be expressed and negotiated among the collective. Secondly, the materials that they collected where always dependent on the encounter with "the other" that required mutual respect and trust.

Figure 9B.4. *Scenic Experience Foz Afora* (*Source*: Cicero Rodrigues, 2017)

Figure 9B.5. *Exhibition Foz Afora* (*Source*: Cicero Rodrigues, 2017)

Then, these materials were shared and discussed for the creation of artistic experimentations that were developed together with the members of the communities. Finally, when the *Coletivo Líquida Ação* brought the multiple facets of their research to the theatre and gallery in Rio de Janeiro, they invited other professionals and local people from Regência to participate in round tables to discuss the scientific and sociocultural situation in the region and open the debate to a larger public. The round tables were composed by "invited researchers, community leaders and/or experts from different areas who study and/or experience socio-environmental impacts caused by mega-mining activities, [they] propose[d] debates and reflections on local realities and global problems." (Rodrigues, 2017) (Figure 9B.6).

The events at the cultural space in Rio de Janeiro comprised the round tables, the artistic exhibition with several installations, talks with different professionals, a concert and scenic experimentations with the participation of artists of a local theatre group of Regência, who were also contemplated by an artistic fund from the state of Espírito Santo and were therefore able to participate in the events. These activities served as a form

Figure 9B.6. *Round Table Discussions (Source*: Coletivo Liquida Acao, 2018)

of closure for this part of the project. However, as Eloisa Brantes emphasised this was not the end of the project. It is an open-ended process that has shed light on the multiple potentials that can evolve from such a research. The collective is already working on a new proposal to take this research to the next level (Mendes, 2018).

9B.4. Conclusion

As exposed in this chapter, arts-based research and environmental aesthetics can make an enriching contribution to the debates on transdisciplinary sustainability. The example of the work of the *Coletivo Líquida Ação* and their projects *Volume Morto* and *Foz Afora* indicate the richness that an artistic research can bring to approaches to environmental and social issues.

The article has introduced the concept of transdisciplinarity and its relation to sustainability. It indicated that transdisciplinary approaches to sustainability propose opening its debates to engage practical actions with theory, following however a non-hegemonic set of ethical steps. Transdisciplinarity has been argued to facilitate the understanding that sustainability is processual and not a stagnated goal to be reached. It also indicates that what sustainability means differs according to sociocultural and environmental circumstances. There are different levels of reality and of perception of reality (Nicolescu, 2006) therefore there cannot be one universal model for sustainability. In this sense, environmental aesthetics and arts-based research were introduced as enhancing different forms of knowledge and as potential strategies to explore alternatives for a transdisciplinary sustainability.

The text followed then with the presentation of the Brazilian artistic collective *Coletivo Líquida Ação* and how the group engages in artistic research about the concept and substance of water. It briefly described the project *Volume Morto* which was the precursor of the project *Foz Afora* (analysed in more detail later in the text). In this description it indicated how current social and environmental circumstances also influenced the process and design of the group's research.

Then the case of the burst of the Brazilian mining dam, *Barragem do Fundão*, was very shortly introduced to contextualise the analysis of the project *Foz Afora*. The aim of the chapter was not to go deep into analysing this environmental accident/crime, nor to investigate the neglect of the companies *Samarco*, *BHP Billiton*, and *Vale* towards their environmental and social responsibility. Some details were briefly exposed, and further literature is indicated in the references.

Finally, the project *Foz Afora* was presented and analysed considering the sociocultural and environmental circumstances faced by the artist-researchers of the *Coletivo Líquida Ação*. However, more significantly, the chapter focuses on the characteristics of their artistic immersion in this environment and how this form of research may contribute to a transdisciplinary sustainability. The analysis indicated how the artist approached the communities affected by the failure of the dam and what kind of strategies they used to collect the material for their artistic experimentations. It highlighted the immersive and non-hierarchical qualities of their approach, as well as the multifaceted dimension of their work. The collective works with intensive corporeal research investigating the relations of oral narratives with bodily memory and spatial perception, which has proven to be very powerful to avoid imposing discourses or pre-defined expectations on their process of research.

Lastly, I would like to emphasise that the aim of the chapter was not to disregard scientific research, nor to replace transdisciplinarity with arts-based research. The argument relies on accepting artistic practice as a valid form of research that may bring fruitful outcomes to rethink sustainability (from a transdisciplinary point of view). Furthermore, another part of the argument refers to overcoming the limited understanding of environmental aesthetics as a form to measure the superficial qualities of a landscape. Environmental aesthetics has a much denser value, since it permeates all dimensions of multi-sensorial and environmental perception. As demonstrated in the analysis of this chapter, it can be enhanced by artistic approaches, but it is not limited to artistic practice or appreciation. Thus, it can be very insightful in the process of rethinking sustainability, mainly in relation to transdisciplinarity.

Chapter 10

A REFLECTION

Frank Birkin[*,‡] and Thomas Polesie[†,§]

University of Sheffield, Sheffield S10 2TN, UK
†*University of Gothenburg, Vasagatan 1, 405 30 Gothenburg, Sweden*

‡*f.birkin@sheffield.ac.uk*
§*Thomas.polesie@handels.gu.se*

The authors of the preceding chapters speak well enough for themselves in their own voices, from all quarters of the world, with their own perspectives, and yet, we argue, they are united. They are united because they all see the world, and work within it, with the same possibility of knowledge.

For Foucault (2002 [1966]), episteme change explained three differences in our approaches to knowledge in Europe from Renaissance times to the Modern day. These changes defined different ages. Birkin and Polesie (2011) used Foucault's explanations to explain a transition within which they live. The Modern possibility of knowledge is for Birkin and Polesie (2011) being replaced by something new in the Human Sciences: the knowledge derived from the Modern abstract, rational belief systems such as neoclassical economic theory, traditional accounting and free markets is being replaced by pragmatic and empirical knowledge based on observations and experience of the world in which we live. It is a new possibility based on the best knowledge we have of how the world, including ourselves, works — science — and some parts of those long-term experiments in living — well that we call traditions.

This book, these authors and many more people worldwide in all sectors of knowledge and society are living and working in the emerging episteme. This new episteme was named "Primal" by Birkin and Polesie (2011) because of its reliance upon our new knowledge of man's origins. All of us are not creating this new episteme — technically that is happening as epistemological man, as well as "his footprints in the sand" (Foucault 2002[1966]), are being washed away by new knowledge. What this book, these authors and many more people worldwide are doing is living through, working within and recording the *consequences* of the emerging episteme — and these are many and varied — think of it as entering a whole new world.

A new possibility of knowledge requires that we reflect on all that we know. In this case, the transition from the Modern to the Primal episteme, only knowledge in the human sciences is affected. But in the past, all of knowledge in Europe has been affected (Foucault, 2002[1966]). This is a big task that takes decades to work out and for some it does not happen at all.

This book is to help clarify and facilitate the transition for we urgently need to think how we live, develop and succeed on our planet. This too is a big task since so much is affected but there is a simple point of access — you, your thoughts, attitudes, values and interpretations. This is the time for new beginnings, startling insights, eureka moments, revelations, personal journeys and transformations, and jokes.

Some 20 years ago, the European Accounting Congress held its annual convention in the beautiful city of Bergen in Norway. The opening ceremony featured an excellent young Norwegian student band and after their music one of their members stepped forwards told the following joke.

> One day, the Norwegian cabinet met to review their nation's capability for defence against any kind of military attack. Ministers for the air force, the army, the navy and the economy sat around the table. The Prime Minister started by asking the minister responsible for the air force to explain how her forces would repel any invasion. After some three hours of listening and debating the review of Norwegian air power was completed.

Faces around the table were anxious; all their planes, helicopters and missiles seemed so grossly inadequate in face of the challenge they might one day have to face as a nation. So too in turn did the ministers for the army and the navy review their many forces and did little to reassure the assembled ministers of state. It was then very late of the day for each review had taken several hours to complete. Wearily the meeting was about to conclude that even with all their combined forces there was little that Norway could do to repel a determined attack. Sadly with heavy hearts they shuffled papers and prepared to leave the meeting.

"Hold on", said the minister responsible for the economy. "Do not forget the work we economists do! You should let we economists have unbridled power and within only a few years we will have our job complete and at that time there will be nothing left in Norway of any value. Nobody will want to invade Norway!"

The assembly of accountants at the European Congress laughed. It was, after all, a joke about economists.

There is some truth of course in the joke. It would not work otherwise. And that truth is now being written large enough to deter invaders from outer space. Our planet, its people and its wildlife are suffering and we all urgently need to change our ways. Indeed some economic thinking may be misleading but the root cause of our problems is best expressed as a *mistake of civilisation* for which so many of us are responsible.

So here we are in Chapter 10 at the end of a book seeking to change a civilisation. This can seem so grand, so ambitious, so exalted, so spirited, so young, so inexperienced, so naïve and so ridiculous.

I am nearing my seventieth birthday as I write so young I am not! Neither is the Chinese civilisation with some 3,000 years of existence and the Chinese are right now changing their civilisation. Also, few who know the Chinese would ever describe them as naïve. Is it "inexperienced"? I think not for the essence of the emerging Primal episteme is that it is based on the formal experiences of exploratory science as well as those of ancient traditions. Neither is it grand if it comes to us in a simple way when we wake up one day to see afresh the world around us. It is ambitious, perhaps magnificently so, but only in aggregate since on a personal level it is about people in daily lives but with new knowledge, insights and understandings. It is a spirited project and not something cold and rational. It has to be spirited: it has to be life giving. We urgently need a new spirit for development and engaging with each other and with the world.

Maybe if our views gain credence, they will consolidate and become doctrinaire like so much of that which people use to lean upon: but perhaps not. Perhaps at this unique time in human history, we can learn to maintain open and inquisitive minds aware of the world in its entirety and neither dependent upon nor closed by the office, the bank and the shopping mall. Perhaps we can all become a lot more selfish — far more selfish than any terrorist or billionaire — and recognise what we really are — inside and out, upside down and right way up, far and near, have been and what will be, knowing little and knowing a lot, in gardens and cities, in sea and on mountain, on Earth and out in the solar system and so on. For a joined-up world is emerging — united by reality.

World I-kuan Tao Headquarters was established in Los Angeles in 1996. I-kuan Tao is short for "The True inheritance of the Original Great Tao." Confucius said of the Tao: "The Tao I follow is the one that unifies all." When asked what this means, a disciple of Confucius answered: "The doctrine of our teacher is to be true to the principles of our nature and the benevolent practice to others."

I grew up in a beautiful Andalusian Valley, hidden in the last foothills of the Sierra Nevada mountain range, on the sea shore, in front of the African coast. In that Moorish land, in my unforgettable Rodaquilar, my spirit was freely formed and my body developed. No one spoke to me about God or the laws; and I made my own laws and skipped God. There I felt the adoration of pantheism, the raw craving for noble things, the repugnance for lies and conventions. I spent my teenage years as a daughter of nature, dreaming with a book in my hand at the seashore or galloping across the mountains…"

Carmen de Burgos (1867–1932)

Dead man naked they shall be one

With the man in the wind and the west moon;

When their bones are picked clean and the clean bones gone,

They shall have stars at elbow and foot;

Though they go mad they shall be sane,

Though they sink through the sea they shall rise again;

Though lovers be lost love shall not;

And death shall have no dominion.

Dylan Thomas (1919–1953)
"And death shall have no dominion."

References

Abbott, A. (2010). Varieties of Ignorance. *The American Sociologist*, *41*(2) (June), pp. 174–189.

Adams, C.A. (2004). The Ethical, Social and Environmental Reporting-Performance Portrayal Gap. *Accounting, Auditing and Accountability Journal*, *17*(5), pp. 731–757.

Adams, M. (2016). #Hypernormaisation- and Why Heathrow Plan is Proof we Exist in a Catastrophic Fantasyland. The Conversation https://theconversation.com/hypernormalisation-and-why-heathrow-plan-is-proof-we-exist-in-a-catastrophic-fantasyland-67674

Ahluwalia, P. (2001). *Politics and Post-colonial Theory: African Inflections*. London: Routledge.

Ahmed, W. (2011). From Mixed Economy to Neoliberalism: Class and Caste in India's Policy Transition. In W. Ahmed, A. Kundu and R. Peet (eds.), *India's New Economic Policy: A Critical Analysis*. London: Routledge, pp. 33–56.

Ajzen, I. and Fishbein, M. (1977). Attitude-behavior Relations: A Theoretical Analysis and Review of Empirical Research. *Psychological Bulletin*, *84*(5), p. 888.

Ajzen, I. and Fishbein, M. (1980). *Understanding Attitudes and Predicting Social Behavior*. Englewood Cliffs: Prentice-Hall.

Altieri, M.A. (1983). The Question of Small farm Development: Who Teaches Whom? *Agriculture, Ecosystems & Environment*, *9*, pp. 401–405.

Altieri, M.A. (1993). Ethnoscience and Biodiversity: Key Elements in the Design of Sustainable Pest Management Systems for Small Farmers in Developing Countries. *Agriculture, Ecosystems & Environment*, *46*, pp. 257–272.

Alvesson, M. and Spicer, A. (2012). A Stupidity-based Theory of Organizations. *Journal of Management Studies, 49*(7), pp. 1194–1220.

Ammi, C. (2013). *Global Consumer Behaviour*. London: John Wiley & Sons.

Anand, G. and Chang, A. (2010). Green Revolution in India Wilts as Subsidies Backfire. The Wall Street Journal [Online], 22 February. Retrieved from http://online.wsj.com/news/articles/SB1000142405274870361590457505292 1612723844

Architects Forum Kenya. (2015). City Council of Nairobi: A Guide of Nairobi City Developments Ordinances and Zones. Retrieved from https://architects-forumkenya.files.wordpress.com/2015/02/zoning-of-nairobi.pdf

Argyris, C. (1993). *Knowledge for Action: A Guide to Overcoming Barriers to Organizational Change*. San Francisco, CA: Jossey-Bass.

Argyris, C. (2004). Reflection and Beyond in Research on Organizational Learning. *Management Learning, 35*(4), pp. 507–509.

Argyris, C. and Schon, D.A. (1978). *Organizational Learning: A Theory of Action Perspective*. Reading, MA: Addison-Wesley.

Arrázola, R. Palenque (1970). *Secretos de la Historia de Cartagena*. Cartagena: Ediciones Hernández.

Australian Curriculum, Assessment and Reporting Authority (2016). Cross-Curriculum Priorities. Retrieved from http://www.acara.edu.au/curriculum/cross-curriculum-priorities

Australian Government (2017, May 8). Early Years Learning Frame-work. Retrieved from https://www.education.gov.au/early-years-learning-framework

Balanzo, A. (2016). Unfolding Capacity: Strategies of Farmers' Organizations as Change Agents. (PhD), University of Twente, The Netherlands. Retrieved from https://doi.org/10.3990/1.9789036541862

Banerjee, S.B. (2002). Corporate Environmentalism: The Construct and its Measurement. *Journal of Business Research, 55*, pp. 177–191.

Banerjee, S.B., Chio, V. and Mir, R. (2009). Organizations, *Markets and Imperial Formations: Towards an Anthropology of Globalization*. Cheltenham: Edward Elgar.

Bansal, P. and Roth, K. (2000). Why Companies Go Green: A Model of Ecological Responsiveness. *Academy of Management Journal, 43*(4), pp. 717–736.

Barbier, E.B. and Burgess, J.C. (2017). Natural Resource Economics, Planetary Boundaries and Strong Sustainability. *Sustainability, 9*(10), article 1858.

Barone, T. and Eisner, E.W. (1997). Arts-based educational research. In: Jaeger, R.M. (ed.). *Complementary Methods for Research in Education*. Washington: American Educational Research Association, pp. 73–116.

Barthes, R. (1971). *Sade, Fourier, Loyola*. Paris: Editions de Seuil. (English trans. by Farrar, Strauss, and Giroux. Baltimore, MD: John Hopkins University Press. 1976)

Bebbington, K.J. (2007). *Accounting for Sustainable Development Performance*. London: CIMA.

Bebbington, K.J. and Gray, R.H. (2001). An Account of Sustainability: Failure, Success and a Reconception. *Critical Perspectives on Accounting*, *12*(5), pp. 557–587.

Bebbington, K.J., Gray, R.H. Hibbitt, C. and Kirk, E. (2001). *Full Cost Accounting*: *An Agenda for Action*, London: ACCA.

Bebbington, K.J. and Unerman, J (forthcoming). Achieving the United Nations Sustainable Development Goals: An Enabling Role for Accounting Research. *Accounting, Auditing and Accountability Journal*.

Bedinelli, T. (2017). "Samarco pagou só 1% do valor de multas ambientais por tragédia de Mariana." El País. Retrieved from: https://brasil.elpais.com/brasil/2017/08/08/politica/1502229456_738687.html

Belz, F.M. and Peattie, K. (2009). *Sustainability Marketing: A Global Perspective*. Chichester: Wiley.

Bentley, T., Bound, K. and Skidmore, P. (2005). World Class — How Sweden Could Reinvent the Rules of Political Economy Over the Next Generation. Confederation of Swedish Enterprise. Retrieved from https://www.svensktnaringsliv.se/english/publications/world-class-how-sweden-could-reinvent-the-rules-of-political-econ_565751.html

Bergstrand, B.O. and Ullberg-Ornell, P.-E. (2009). *Dynamic Growth Capital — Social Capital and its Importance for Growth and Development*. Lidköping: Social Capital Forum.

Bergstrand, B.-O, Björk F. and Molnar, S. (2011). Biosphere Entrepreneurship. A Pilot Study on Social Entrepreneurship in the Biosphere Reserve Lake Vänern Archipelago and Mount Kinnekulle. Biosfar Vanerskargarden Kinnekulle. Retrieved from http://media.vanerkulle.org/2013/09/297_Biosphere-Entrepreneurship-A-Pilot-Study-Webversion.pdf

Bernstein, Jay, H. (2015). Transdisciplinarity: A Review of its Origins, Development, and Current Issues. *Journal of Research Practice*, *11*(1), Retrieved from: http://jrp.icaap.org/index.php/jrp/article/view/510/412

Berthrong, J. (2003). Confucian Views of Nature. In H. Selin (ed.). *Nature Across Cultures: Views of Nature and the Environment in Non-western Cultures* (pp. 373–392). London: Kluwer Academic Publishers.

Bhabha, H. (2004). *The Location of Culture — 1994*. London and New York: Routledge.

Bhaskar, R. (1978). *A Realist Theory of Science*. Brighton: Harvester Press.

Birkin, F.K. (1996). The Ecological Accountant: From Cognito to Thinking Like a Mountain. *Critical Perspectives on Accounting, 7*(3), pp. 231–257.

Birkin, F.K. and Polesie, T. (2011). An Epistemic Analysis of (un)sustainable Business. *Journal of Business Ethics, 103*, pp. 239–253.

Birkin, F. K. and Polesie, T. (2012). *Intrinsic Sustainable Development: Epistemes, Science, Business and Sustainability*, Singapore: World Scientific Press.

Biswas, W.K. (2012). The Importance of Industrial Ecology in Engineering Education for Sustainable Development. *International Journal of Sustainability in Higher Education, 13*(2), pp. 119–132.

Bloomburg, M. (2017). Michael Bloomberg on President Trump, Brexit, and Climate Change. Retrieved from http://fortune.com/2017/10/13/michael-bloomberg-trump-climate-change/

Bond, M.H. (1996). *The Handbook of Chinese Psychology*. Hong Kong: Oxford University Press.

Bonney, C. and Sternberg, R. J. (2011). Learning to Think Critically. In Mayer R. E., and Alexander P. A. (eds.), *Handbook of Research on Learning and Instruction* (pp. 166–195). New York: Routledge.

Borland, H. (2009a). Conceptualising Global Strategic Sustainability and Corporate Transformational Change. *International Marketing Review, 26*(4/5), pp. 554–572.

Borland, H. (2009b). Definitions, Theories, Drivers and Managerial Implications: Grounding Global Strategic Sustainability. *Academy of Marketing Science Conference*, May, Baltimore.

Borland, H., Ambrosini, V., Lindgreen, A. and Vanhamme, J. (2016). Building Theory at the Intersection of Ecological Sustainability and Strategic Management. *Journal of Business Ethics, 135*(2), pp. 293–307.

Bosselmann, K. (1995). *When Two Worlds Collide: Society and Ecology*. Auckland: RSVP Publishing.

Boud, D., Cohen, R. and Walker, D. (1993). *Using Experience for Learning*. Buckingham: Open University Press.

Brandt, P., Ernst, A., Gralla, F., Luederitz, C. and von Wehrden, H. (2013). A Review of Transdisciplinary Research in Sustainability Science. *Ecological Economics, 92*, pp. 1–15.

Brante, E. (2017). Na Janela do Trem. Coletivo Líquida Ação. Retrieved from: https://www.coletivoliquidaacao.com/single-post/2017/06/15/Na-janela-do-trem

Brattberg, G. (1995). To Meet Long-lasting Pain. Stockholm.

Brookfield, S.D. (1995). *Becoming a Critically Reflective Teacher*. San Francisco: Jossey-Bass Publishers.

Brown, L.R. and Flavin, C. (with others) (1999). State of the World 1999, London: Earthscan/Worldwatch Institute.

Buhr, N., Gray, R. and Milne, M. (2014). Histories, Rationales, Voluntary Standards and Future Prospects or Sustainability Reporting: CSR, GRI, IIRC and Beyond. In *Sustainability Accounting and Accountability* (eds.), J. Bebbington J. Unerman, and B. O'Dwyer (pp. 51–71). London: Routledge.

Bureau, (2016). Govt Team Leaves for Mozambique to Explore Pulses Import. The Hindu [Online], 21 June. Retrieved from: http://www.thehindubusinessline.com/economy/agri-business/govt-team-leaves-for-mozambique-to-explore-pulses-imports/article8756040.ece

Burgess, J., Harrison, C.M. and Filius, P. (1998). Environmental Communication and the Cultural Politics of Environmental Citizenship. *Environment and Planning, 30*(8), pp. 1445–1460.

CafCu Media (2015). School Lunch in Japan — It's not Just About Eating! [Video File]. Retrieved from https://www.youtube.com/watch?v=hL5mKE4e4uU

Calderhead, J. (1989). Reflective Teaching and Teacher Education. *Teaching and Teacher Education, 5*(1), pp. 43–51.

Capra, F. (2004). The hidden connections: A science for sustainable living. New York: Anchor Books.

Caradonna, J.L. (2014). *Sustainability: A History.* New York: Oxford University Press.

Carmo, L.F., Kamino, L.C.Y., Tobias Junior, R., de Campos, C. and Pinto, C.E.F. (2017). Fundão Tailings Dam Failures: The Environment Tragedy of the Largest Technological Disaster of Brazilian Mining in Global Context. *Perspectives in Ecology and Conservation, 15*, pp. 145–151.

Carson, R. (2002). *Silent Spring.* Boston: Mariner Books.

CEPAC. (2003). Historia del Pueblo Afrocolombiano — Perspectiva Pastoral. Retrieved from http://axe-cali.tripod.com/cepac/hispafrocol/6.htm

Cervantes, G. (2007). A Methodology for Teaching Industrial Ecology. *International Journal of Sustainability in Higher Education, 8*(2), pp. 131–141.

Chadha, S. (2017). VUCA world: Provoking the future. *Human Capital, 20*(8), pp. 14–18.

Chan, K.M. (2014). Harmonious Society. International Encyclopedia of Civil Society. Retrieved from https://www.cuhk.ecu.hk/cenmtre/ccss/publications/km....../CKM_14.pdf

Chan, R.Y.K. (1999). Environmental Attitudes and Behaviour of Consumers in China. *Journal of International Consumer Marketing, 11*(4), pp. 25–52.

Chandrasekhar, C.P. (2013). Locked into Business. Frontline. [Online], July 10. Retrieved from: http://www.frontline.in/cover-story/locked-into-business/article4888071.ece

Cheang, C.C., So, W.-M., Zhan, W.Y. and Tsoi, K.H. (2017). Education for Sustainability Using a Campus Eco-garden as a Learning Environment. *International Journal of Sustainability in Higher Education*, *18*(2), pp. 242–262.

Cheng, T.-M. and Wu, H.C. (2015). How do Environmental Knowledge, Environmental Sensitivity, and Place Attachment Affect Environmentally Responsible Behavior? An Integrated Approach for Sustainable Island Tourism. *Journal of Sustainable Tourism*, *23*(4), pp. 557–576.

Child, J., Faulkner, D. and Tallman, S. (2005). *Cooperative Strategy*. Oxford: Oxford University Press.

Choudhury, N. (1988). The Seeking of Accounting where it is not: Towards a Theory of Non-accounting in Organizational Settings. *Accounting Organizations and Society*, *13*(6), pp. 549–557.

CIA. (2017). *The World Fact Book*. Washington: CIA.

Clarke, I., Flaherty, T.B. and Mottner, S. (2001). Student Perceptions of Educational Technology Tools. *Journal of Marketing Education*, *23*, pp. 169–177.

Clayre, A. (1976). *The Heart of the Dragon*. London: Collins Harvill.

Closs, D., Speier, C. and Meacham, N. (2011). Sustainability to Support End-to-End Value Chains: The Role of Supply Chain Management. *Journal of the Academy of Marketing Science*, *39*(1), pp. 101–116.

Cole, M.S., Feild, H.S. and Harris, S.G. (2004). Student Learning Motivation and Psychological Hardiness: Interactive Effects on Students' Reactions to a Management Class. *Academy of Management Learning & Education*, *3*, pp. 64–85.

Coletivo Líquida Ação (2011). Mudamos Novo Site. Retrieved from: http://coletivoliquidaacao.blogspot.com.br/2010/08/ouro-liquido-mitolorgia-urbanas.html

Coletivo Liquida Acao (2018). Coletivo Liquida Acao. Retrieved from: https://www.coletivoliquidaacao.com/

Collaborative Economy. (2017). Collaborative Economy: Growth. Retrieved from: http://ec.europa.eu/growth/single-market/services/collaborative-economy_en

Collison, D., Dey, C. Hannah, G. and Stevenson, L. (2007). Income Inequality and Child Mortality in Wealthy Nations. *Journal of Public Health*, *29*(2), pp. 114–117.

Collison, D., Dey, C. Hannah, G. and Stevenson, L. (2010). Anglo-American Capitalism: The Role and Potential of Social Accounting. *Accounting, Auditing and Accountability Journal, 23*(8), pp. 956–981.

Communist Party of Vietnam. (1998). Resolution of the 5th Conference of the Central Committee and the Communist Party of Vietnam on Building a Progressive Culture, Imbued with National Identity. Communist Party of Vietnam, Number 03/NQ-T.

Cook, K., Brown, A. and Ballard, G. (2016). Using Photovoice to Explore Environmental Sustainability Across Languages and Cultures. *Discourse and Communication for Sustainable Education, 7*(1), pp. 49–67.

Cooper, C. (1992). The Non and Nom of Accounting for (m)other Nature. *Accounting, Auditing and Accountability Journal, 5*(3), pp. 16–39.

Cooper, C., Taylor, P., Smith, N. and Catchpowle, L. (2005). A Discussion of the Political Potential of Social Accounting. *Critical Perspectives on Accounting, 16*(7), pp. 951–974.

Cooper, J.C. (2010). *The Illustrated Introduction to Taoism: The Wisdom of the Sages*. Bloomington, IN: World Wisdom, Inc.

Corcoran, P.B. and Koshy, K.C. (2010). The Pacific Way: Sustainability in Higher Education in the South Pacific Island Nations. *International Journal of Sustainability in Higher Education, 11*(2), pp. 130–140.

Cotgrove, S. (1982). *Catastrophe or Cornucopia: The Environment, Politics and the Future*. New York: Wiley.

Cotton, D. and Winter, J. (2010). It's Not Just Bits of Paper and Light Bulbs. A Review of Sustainability Pedagogies and Their Potential for use in Higher Education. *Sustainability Education: Perspectives and Practice Across Higher Education*, pp. 39–54.

Cronin, J.J., Smith, J., Gleim, M., Ramirez, E. and Martinez, J. (2011). Green marketing Strategies: An Examination of Stakeholders and the Opportunities they Present. *Journal of the Academy of Marketing Science, 39*(1), pp. 158–174.

Cummings, S. and Angwin, D. (2015). *Strategy Builder: How to Create and Communicate More Effective Strategies*. Chichester: Wiley.

Curry, P. (2011). *Ecological Ethics: An introduction* (2nd ed.). Chichester: Wiley.

Curtis, D.J., Reid, N. and Reeve, I. (2014). Towards Ecological Sustainability: Observations on the Role of the Arts. *S.A.P.I.E.N.S 7*(1), pp. 1–15.

DANE. (S.F.). La Visibilidad Estadística de los Grupos Étnicos Colombianos. Bogotá: Imprenta Nacional República de Colombia.

Dao, D.A. (1939). *The Fundamentals of Vietnamese History and Culture*. Hanooi, Vietnam, Van Hoa Thong Tin Publisher.

Dao, D.A. (1957). Chinese Vietnamese Dictionary (3e), Hanoi, Vietnam: Van Hoa Truong Thi Publisher.

Dashwood, H.S. and Puplampu, B.B. (2010). Corporate Social Responsibility and Canadian Mining Companies in the Developing World: The Role of Organizational Leadership and Learning. *Canadian Journal of Development Studies/Revue Canadienne d'études du Dévelopement, 30*, pp. 175–196.

Daviron, B. and Ponte, S., (2005). *The Coffee Paradox: Global Markets, Commodity Trade and the Elusive Promise of Development*. London: Zed Books in Association with the CTA.

Davison, A. (2001). *Technology and the Contested Meanings of Sustainability*. New York: State University of New York Press.

Davison, J. and Warren, S. (2009). Imag(in)ing Accounting and Accountability. *Accounting, Auditing and Accountability Journal, 22*(6), pp. 845–857.

Deegan C. (2016). Twenty Five Years of Social and Environmental Accounting Research Within Critical Perspectives of Accounting: Hits, Misses and Ways Forward. *Critical Perspectives on Accounting, 43*, pp. 65–87.

de Friedemann, N.S. (1987). Ma Ngombe: Guerreros y Ganaderos en Palenque: C. Bogota: Valencia Editores.

de Groot, J.I.M. and Steg, L. (2007). Value Orientations and Environmental Beliefs in Five Countries: Validity of an Instrument to Measure Egoistic, Altruistic and Biospheric Value Orientation. *Journal of Cross-Cultural Psychology, 38*(3), pp. 318–332.

Deshpande, R.S. and Arora, S. (2010). Editors' Introduction. In *Agrarian Crisis and Farmers' Suicides*, (eds. R.S. Deshpande, and S. Arora). Thousand Oaks, CA: SAGE Publications.

Dewey, J. (1933). *How We Think*. New York: Cosimo, Inc.

Dey, C. (2007). Social Accounting at Traidcraft Plc: A Struggle for the Meaning of Fair Trade. *Accounting, Auditing and Accountability Journal, 20*(3), pp. 423–445.

Dey, C., Russell, S. and Thomson, I. (2011). Exploring the Potential of Shadow Accounts in Problematising Institutional Conduct. In Osbourne, S. and A. Ball (eds.), *Social Accounting and Public Management: Accountability for the Common Good* (pp. 64–75). Abingdon: Routledge.

Diamond, J. (2006). *Collapse: How Societies Choose to Fail or Succeed*. London, Penguin.

Dieleman, H. (2012). Transdisciplinary Artful Doing in Spaces of Experimentation and Imagination. *Transdisciplinary Journal of Engineering & Science, 3*, pp. 4–57.

Diesendorf, M. and Hamilton, C. (eds.) (1997). *Human Ecology, Human Economy: Ideas Towards An Ecologically Sustainable Future.* Sydney: Allen and Unwin.

Dornelas, R.S. *et al.* (2016). Ações Civis Públicas e Termos de Ajustamento de Conduta no Caso do Desastre Ambiental da Samarco: Considerações a Partir do Observatório de Ações Judiciais. In: M., Bruno and Cristina Losekann (eds.). Desastre no Vale do Rio Doce: Antecedentes, Impactos e Ações Sobre a Destruição. Rio de Janeiro: Folio Digital: Letra e Imagem. pp. 339–369.

Dostoyevsky, F. (1984). Dnevnik Pisatelya. In A Writer's Diary. 1876. July and August, pp. 252–256. Sankt — Perersburg: Nauka.

Dryzek, J. (1997). *The Politics of the Earth: Environmental Discourses.* Oxford: Oxford University Press.

Du Nann Winter, D. and Koger, S. (2004). *The Psychology of Environmental Problems* (2nd ed.). Mahwah, NJ: Lawrence Erlbaum Associates/Eurospan (London).

Dunlap, R., Van Liere, K., Mertig, A. and Jones, R. (2000). New Trends in Measuring Environmental Attitudes: Measuring Endorsement of the New Ecological Paradigm: A Revised NEP Scale. *Journal of Social Issues, 56*(Fall), pp. 425–442.

Durr, E., Bilecki, J. and Li, E. (2017). Are Beliefs in the Importance of Pro-environmental Behaviors Correlated with Pro-environmental Behaviors at a College Campus? *Sustainability: The Journal of Record, 10*(3), pp. 204–210.

Dussel, E. (2000). Europa, Modernidad y Eurocentrismo. In E. Lander and S. Castro-Gómez (eds.), *La Colonialidad Del Saber: Eurocentrismo y Ciencias Sociales.* Buenos Aires: Clacso.

Du, Y. and Gray, R. (2013). The Emergence of Stand Alone Social and Environmental Reporting in Mainland China: An Exploratory Research Note. *Social and Environmental Accounting Journal, 33*(2), pp. 104–112.

Dyment, J.E., Davis, J.M., Nailon, D., Emery, S., Getenet, S., McCrea, N. and Hill, A. (2014). The Impact of Professional Development on Early Childhood Educators' Confidence, Understanding and Knowledge of Education for Sustainability. *Environmental Education Research, 20*(5), pp. 660–679.

Dyment, J. E., Hill, A. and Emery, S. (2015). Sustainability as a Cross-curricular Priority in the Australian Curriculum: A Tasmanian Investigation. *Environmental Education Research*, *21*(8), pp. 1105–1126.

Eastman, J.K., Iyer, R. and Eastman, K.L. (2011). Business Students' Perceptions, Attitudes, and Satisfaction with Interactive Technology: An Exploratory Study. *Journal of Education for Business*, *86*, pp. 36–43.

Eernstman, N. and Arjen E.J.W. (2013). Locative Meaning making: An Arts-based Approach to Learning for Sustainable Development. *Sustainability*, *5*(4), pp. 1645–1660.

Ekins, P. (2000). *Economic Growth and Environmental Sustainability*. London: Routledge.

Ellis, N., Fitchett, J., Higgins, M., Jack, G., Lim, M., Saren, M. and Tadajewski, M. (2011). *Marketing: A Critical Textbook*. London: SAGE Publications.

Elumelu. (2018). What is Africapitalism? The Tony Elumelu Foundation Anticapitalism Institute. Retrieved from: http://tonyelumelufoundation.org/africapitalisminstitute/about-us/what-is-africapitalism/

Erkman, S. (1997). Industrial Ecology: An Historical View. *Journal of Cleaner Production*, 5 (1–2), pp. 1–10.

Evans, T. L. (2015). Transdisciplinary Collaborations for Sustainability Education: Institutional and Intragroup Challenges and Opportunities. *Policy Futures in Education*, *13*(1), pp. 70–96.

Fals-Borda, O. and Rahman, M.A. (1991). *Action and Knowledge: Breaking the Monopoly with Participatory Action Research*. New York: Apex Press.

Fanon, F. (2008). *Black Skin, White Masks*. London: Grove Press.

Fanon, F. and Farrington, C. (1969). The Wretched of the Earth. Translated by Constance Farrington. (Reprint). London: Penguin Books.

Fantz, A. (2008). "Mekong, a 'Treasure Trove' of 1,000 Newly Discovered Species," CNN, December 16.

Farias, G., Farias, C. M. and Fairfield, K. D. 2010. Teacher as Judge or Partner: The Dilemma of Grades Versus Learning. *Journal of Education for Business*, *85*, pp. 336–342.

Feger, C. and Mermet, L. (2017). A Blueprint Towards Accounting for the Management of Ecosystems. *Accounting, Auditing & Accountability Journal*, *30*(7), pp. 1511–1536.

Fernandes, G.O., Goulart, F.F., Ranieri, B.D., Coelho, M.S. and Soares-Filho, B. (2016). Deep into the Mud: Ecological and Socio-economic Impacts of the Dam Breach in Mariana, Brazil. *Natureza & Conservação*, *14*, pp. 35–45.

Filipovic, V. and Ullberg Ornell, P-E. (2003). Dream City Lidköping — Sarajevo. A Pilot Project About Development of Local Democracy. Lidköping: Lidköping Kommun.

Foucault, M. (2002 [1966]). *The Order of Things — An Archaeology of the Human Sciences*. London: Routledge Classics.

Foz Afora. (2017). O Projecto Foz Afora. Coletivo Líquida Ação. Retrieved from: https://www.coletivoliquidaacao.com/projetofozafora.

Franklin, A. (2002). *Nature and social theory*. London: SAGE Publications.

Freire, P. (2000). *Pedagogy of the Oppressed*. New York: Bloomsbury Publishing.

Frenkel, M. (2008). The Multinational Corporation as a Third Space: Rethinking International Management Discourse on Knowledge Transfer Through Homi Bhabha. *Academy of Management Review*, *33*, pp. 924–942.

Freund, B. (2007). *The African City: A history*. Leiden: Cambridge University Press.

Gaia, S. and Jones, M.J. (2017). UK Local Councils Reporting of Biodiversity Values: A Stakeholder Perspective. *Accounting, Auditing & Accountability Journal*, *30*(7), pp. 1614–1638.

Gallhofer S., Haslam, J., Monk, E. and Roberts, C. (2006). The Emancipatory Potential of Online Reporting: The Case of Counter Accounting. *Accounting, Auditing and Accountability Journal*, *19*(5), pp. 681–718.

Gao, Y. (2009). Corporate Social Performance in China: Evidence from Large Companies. *Journal of Business Ethics*, *89*, pp. 23–35.

Garvin, T., McGee, T.K., Smoyer-Tomic, K.E. and Aubynn, E.A. (2009). Community–company Relations in Gold Mining in Ghana. *Journal of Environmental Management, 90*, pp. 571–586.

General Statistics Office. (2017). *Statistical Yearbook of Vietnam 2016*. Hanoi, Vietnam: Statistical Publishing House.

Gerstner, A. (2011). Leadership and Organizational Patterns in the Daodejing. *The Journal of Management Development*, *30* (7/8), pp. 675–684.

Gibson K., Gray, R. Laing, Y. and Dey, C. (2001). *The Silent Accounts Project: Draft Silent and Shadow Accounts Tesco Plc 1999–2000*. Glasgow: CSEAR.

Giddens, A. (1991). *Modernity and Self-identity*. Cambridge: Polity Press.

Gieryn, T.F. (1983). Boundary-work and the Demarcation of Science from Non-science: Strains and Interests in Professional Ideologies of Scientists. *American Sociological Review*, pp. 781–795.

Gladwin, T., Kennelly, J. and Krause, T.S. (1995). Shifting Paradigms for Sustainable Development: Implications for Management Theory and Research. *Academy of Management Review*, *20*(4), pp. 874–907.

Gleick, J. (1988). *Chaos — Making a New Science*. London: Cardinal Sphere Books.

Gliessman, S.R. (2007). *Agroecology: The Ecology of Sustainable Food Systems*, (2nd ed.). Boca Raton, FL/London: CRC.

Global Footprint Network. (2017). Making the Sustainable Development Goals Consistent with Sustainability, Retrieved from http://www.footprint-network.org/2017/09/01/making-sustainable-development-goals-consistent-sustainability/

Global Reporting Initiative. (2013). G4 Sustainability Reporting Guidelines. Retrieved from Amsterdam: https://www.globalreporting.org/information/g4/Pages/default.aspx

Goldsmith E. *et al.* (1972). *Blueprint for Survival*. Harmondsworth: Penguin.

Gore, J. M. and Zeichner, K. M. (1991). Action Research and Reflective Teaching in Preservice Teacher Education: A Case Study from the United States. *Teaching and Teacher Education*, *7*(2), pp. 119–136.

Gorz, A. (1989). *Critique of Economic Reason* (*trans G.Handyside and C.Turner*). London: Verso.

Gray, R.H. (1990). The Greening of Accountancy: The Profession After Pearce. London: ACCA.

Gray, R.H. (1992). Accounting and Environmentalism: An Exploration of the Challenge of Gently Accounting for Accountability, Transparency and Sustainability. *Accounting Organisations and Society*, *17*(5), pp. 399–426.

Gray, R. (2006). Social, Environmental, and Sustainability Reporting and Organisational Value Creation? Whose Value? Whose Creation? *Accounting, Auditing and Accountability Journal*, *19*(3), pp. 319–348.

Gray, R. 2010. Is accounting for sustainability actually accounting for sustainability... and how would we know? An exploration of narratives of organisations and the planet. *Accounting, Organizations and Society*, *35*(1), pp. 47–62.

Gray, R.H., Adams, C. and Owen, D.L. (2014). *Accountability, Social Responsibility and Sustainability: Accounting for Society and the Environment*. London: Pearson.

Gray R., Adams, C. and Owen, D. (2017). Social and Environmental Accounting and the Critical Accounting Project(s): In Search of Creative Tension? In R. Roslender (ed.). *The Routledge Companion of Critical Accounting*. London: Routledge, pp. 241–257.

Gray, R.H., Bebbington, K.J. and Walters, D. (1993). *Accounting for the Environment: The Greening of Accountancy Part II*. London: Paul Chapman.

Gray, R., Brennan, A. and Malpas, J. (2014). New Accounts: Towards a Reframing of Social Accounting. *Accounting Forum, 38*(4), pp. 258–273.

Gray, R., Dey, C. Owen, D. Evans, R. and Zadek, S. (1997). Struggling with the Praxis of Social Accounting: Stakeholders, Accountability, Audits and Procedures. *Accounting, Auditing and Accountability Journal 10*(3), pp. 325–364.

Gray, R. and Milne, M. J. (2018). Perhaps the Dodo should have Accounted for Human Beings? Accounts of Humanity and (its) extinction. *Accounting, Auditing and Accountability Journal, 31*(3), pp. 826–848. https://doi.org/10.1108/AAAJ-03-2016-2483

Grönroos, C. (2007). *In Search of a New Logic for Marketing*. Chichester: Wiley.

Guest, R. (2010). The Economics of Sustainability in the Context of Climate Change: An Overview. *Journal of World Business, 45*, pp. 326–335.

Guo, P., Zhong, C., Chen, Y., Wang, X. and Li, W. (2008). A Journey to Discover Values 2008 — Study of Sustainability Reporting in China. Retrieved from Syn Tao (ed.): http://syntao.com/Uploads/%7B065554F3-B9D7-4DDC-8BA9-3DFE894119A9%7D_A%20journey%20to%20discover%20values%202008.pdf

Ha, V.T. (1998). *Tracing Ancient Civilisations*, Hanoi, Vietnam: Hanoi Social Sciences Publisher.

Habermas, J. (1984). *The Theory of Communicative Action: Vol. 1. Reason and the Rationalization of Society* (T. McCarthy, Trans.). Boston: Beacon.

Habermas, J. (1990). *Moral Consciousness and Communicative Action*. Cambridge, MA: MIT Press.

Hall, J. and Vredenburg, H. (2003). The Challenges of Innovating for Sustainable Development. *MIT Sloan Management Review, 45*(1), pp. 61–68.

Halliman, D.M. and Morgan, W.T.W. (1967). The City of Nairobi. In W.T.W. Morgan (ed.), *Nairobi: City and Region* (pp. 98–120). London: Oxford University Press.

Hamari, J., Sjöklint, M. and Ukkonen, A. (2016). The Sharing Economy: Why People Participate in Collaborative Consumption. *Journal of the Association for Information Science and Technology, 67*(9), pp. 2047–2059.

Hammond, G.P. (2004). Engineering Sustainability: Thermodynamics, Energy Systems, and the Environment. *International Journal of Energy Research, 28*, pp. 613–639.

Harris, J. (2013). Can Green Capitalism Build a Sustainable Society? *International Critical Thought, 3*(4), pp. 468–479.

Hart, S. (1995). A Natural-resource-based View of the Firm. *Academy of Management Review*, *20*(4), pp. 986–1014.

Hart, S. (1997). Beyond Greening: Strategies for a Sustainable World. *Harvard Business Review*, *75*(1), pp. 66–76.

Hart, S. (2007). *Capitalism at the Crossroads* (2nd ed.). Englewood Cliffs, NJ: Pearson/Wharton School Publishing.

Hart, S. and Milstein, M. (1999). Global Sustainability and the Creative Destruction of Industries. *MIT Sloan Management Review*, *41*(Fall), pp. 23–34.

Hart, S. and Milstein, M. (2003). Creating Sustainable Value. *Academy of Management Executive*, *17*(2), pp. 56–67.

Harte G. and Owen, D.L. (1987). Fighting De-industrialisation: The Role of Local Government Social Audits. *Accounting, Organizations and Society*, *12*(2), pp. 123–142.

Hartz, P. (2009). *World Religions: Shinto* (3rd ed.). New York: Chelsea House Publishers, p.11.

Hass Consult. (2017). The Property Index: Quarterly Reports. Retrieved from http://www.hassconsult.co.ke/

Hebbar, R. (2010). Framing the Development Debate: The Case of Farmers' Suicide in India. In *Agrarian Crisis and Farmers' Suicides* (R.S. Deshpande and S. Arora, eds.). Thousand Oaks, CA: SAGE Publications.

Heckle, J. and Wals, A.E. (2015). The UN Decade of Education for Sustainable Development: Business as Usual in the End. *Environmental Education Research*, *21*(3), pp. 491–505.

Helfrich, H. (1999). Beyond the Dilemma of Cross-cultural Psychology: Resolving the Tension between Etic and Emic Approaches. *Culture & Psychology*, *5*(2), pp.131–153.

Herbert, J. (1967). *Shintô; at the Fountain-head of Japan*. New York: Stein and Day.

Hill, D. and Kumar, R. (eds.). (2011). *Global Neoliberalism and Education and its Consequences*. New York: Routledge.

Hill, A., Nailon, D., Getenet, S., McCrea, N., Emery, S., Dyment, J. and Davis, J.M. (2014). Exploring How Adults who Work with Young Children Conceptualise Sustainability and Describe their Practice Initiatives. *Australasian Journal of Early Childhood*, *39*(3), p. 14.

Hofstede Insights. (2018). Country Comparison, Hofstede Insights, Retrieved from: https://www.hofstede-insights.com/country-comparison/vietnam/

Hopwood, A., Unerman, J. and Fried, J. (2010). *Accounting for sustainability: Practical insights*. London: Earthscan.

Huenemann, R.W. (2013). Economic Reforms, 1978-present. Retrieved from http://www.oxfordbibliographies.com/view/document/obo-9780199920082/obo-9780199920082-0008.xml

Hue-Tam, H.T. (2017) Religions in Vietnam, Asia Society. Retrieved from: www. asiasociety.org

Hult, G.T.M. (2011). Market-focused Sustainability: Market Orientation Plus! *Journal of the Academy of Marketing Science, 39*(1), pp. 1–6.

IBAMA. (2015). Laudo Técnico Preliminar: Impactos Ambientais Decorrentes do Desastre Envolvendo o Rompimento da Barragem de Fundão, em Mariana, Minas Gerais. Ministerio do Meio Ambiente, Brazil. Retrieved from: http://www. ibama.gov.br/informes/rompimento-da-barragem-de-fundao#autosdeinfracao

IFAC. (2005). International Guidance Document: Environmental Management Accounting. In I.F.O. Accountants (ed.). New York: IFAC.

IFRC. (2017). Viet Nam Red Cross Society Celebrates International Day for Disaster Risk Reduction and Launches the World Disasters Report 2015, International Federation of Red Cross and Red Crescent. Retrieved from: http://www.ifrc.org/en/news-and-media/news-stories/asia-pacific/vietnam/ viet-nam-red-cross-society-celebrates-international-day-for-disaster-risk-reduction-and-launches-the-world-disasters-report-2015-69538/

International Union for Conservation of Nature and Natural Resources (IUCN). (2015). The IUCN Red List of Threatened Species. Retrieved from http:// www.iucnredlist.org/news/2015-iucn-species-highlights

Intrinsic Earth. (2018). Intrinsic Earth. Retrieved from https://www.intrinsicearth. org/

Irwin, R.L. and deCosson, A. (2004). A/r/tography: Rendering self through arts-based living inquiry. Vancouver: Pacific Educational Press.

Iyer, G. (1999). Business, Consumers and Sustainable Living in an Interconnected World: A multilateral Ecocentric Approach. *Journal of Business Ethics, 20*(4), pp. 273–288.

Jack, L. (2007). Accounting, Post-productivism and Corporate Power in UK food and Agriculture. *Critical Perspectives on Accounting, 18*(8), pp. 905–931.

Jacobs, M. (1999). Sustainable Development as a Contested Concept. In A. Dobson (ed.), *Fairness and Futurity: Essays on Environmental Sustainability and Social Justice*. Oxford: Oxford University Press.

Jadhav, R. and Bhardwaj, M. (2017). Centre's Plan to Boost Pulses and Oilseeds Production Becomes Victim of Its Own Success. The Wire. [Online] March 22. Retrieved from: https://thewire.in/agriculture/centres-plan-boost-pulses-oilseeds-production-becomes-victim-success

Jahn, T. (2008). Transdisciplinarity in the Practice of Research. Matthias Bergmann/Engelbert Schramm (eds.). Transdisziplinäre Forschung. *Integrative Forschungsprozesse Verstehen und Bewerten*. Frankfurt/New York: Campus Verlag, pp. 21–37.

Jelinski, L.W., Graedel, T.E., Laudise, R.A., McCall, D.W. and Patel, C.K. (1992). Industrial Ecology: Concepts and Approaches. *Proceedings of the National Academy of Sciences, 89*(3), pp. 793–797.

Jensen, B.B. and Schnack, K. (1997). The Action Competence Approach in Environmental Education. *Environmental Education Research, 3*(2), pp. 163–178.

Jiang, C. (2013). Creating an Ecological Civilization. Qiushi Journal (Organ of the Central Committee of the Communist Party of China, 5(1). Retrieved from http://english.qstheory.cn/magazine/201301/201302/t20130219_211851.htm

Johnson H. Thomas (2107). The Tragedy of Modern Economic Growth: A Call to Business to Radically Change its Purpose and Practices. *Accounting History, 22*(2), pp. 167–178.

Jones M. J. and Solomon, J.F. (2013). Problematising Accounting for Biodiversity. *Accounting, Auditing and Accountability Journal 26*(5), pp. 668–687.

Kagan, S. (2011). Aesthetics of Sustainability: A Transdisciplinary Sensibility for Transformative Practices. *Transdisciplinary Journal of Engineering & Science, 2*, pp. 65–73.

Kamuf, P. (2007). Accounterability. *Textual Practice, 21*(2), pp 251–266.

Karns, G. L. (2005). An Update of Marketing Student Perceptions of Learning Activities: Structure, Preferences, and Effectiveness. *Journal of Marketing Education, 27*, pp. 163–171.

Katakam, A. (2013). It Excludes Farmers, Frontline July 11. Retrieved from: http://www.frontline.in/cover-story/it-excludes-farmers/article4888083.ece

Ketola, T. (2008). A Holistic Corporate Responsibility Model: Integrating Values, Discourses and Actions. *Journal of Business Ethics, 80*, pp. 419–435.

Kenya National Bureau of Statistics [KNBS] (2015a). KNBS Data. Retrieved from http://www.knbs.or.ke

Kenya National Bureau of Statistics [KNBS] (2015b). Census Documents. Retrieved from http://www.knbs.or.ke

Kenya National Bureau of Statistics [KNBS] (2017). Quarterly GDP Reports. Retrieved from https://www.knbs.or.ke/data-releases/

Kilbourne, W.E. (1998). Green Marketing: A Theoretical Perspective. *Journal of Marketing Management, 14*, pp. 641–655.

Kilbourne, W.E. (2008). How Macro should Macromarketing Be? *Journal of Macromarketing, 28*(2), pp. 189–191.

Kilbourne, W.E., Beckman, S.C. and Thelen, E. (2002). The Role of the DSP in Environmental Attitudes: A Multinational Examination. *Journal of Business Research, 55*(3), pp. 193–204.

Killen, R. (2013). *Effective Teaching Strategies: Lessons from Research and Practice*. South Melbourne, Victoria: Cengage Learning Australia.

King, A. (1995). Avoiding Ecological Surprise: Lessons from Long-standing Communities. *Academy of Management Review, 20*(4), pp. 961–985.

Knowles, J.G. and A.L. Cole, A.L. (eds.). (2008). *Handbook of the ARTS in Qualitative Research*. London: SAGE Publications.

Kolb, D.A. (2014). *Experiential Learning: Experience as the Source of Learning and Development*. New Jersey: Pearson Education.

Kollmuss, A. and Agyeman, J. (2002). Mind the Gap: Why do People Act Environmentally and what are the Barriers to Pro-environmental Behavior? *Environmental Education Research, 8*(3), pp. 239–260.

Krishnaraj, M. (2006). Food Security, Agrarian Crisis and Rural Livelihoods: Implications for Women. *Economic and Political Weekly*, pp. 5376–5388.

Kuhn, T.S. (1962). *The Structure of Scientific Revolutions*. Chicago: University of Chicago Press.

Laine, M. (2010). Towards Sustaining the Status Quo: Business Talk of Sustainability in Finnish Corporate Disclosures 1987–2005. *European Accounting Review, 19*(2), pp. 247–274.

Laine, M. and Vinnari, E. (2017). The Transformative Potential of Counter Accounts: A Case Study of Animal Rights Activism. *Accounting, Auditing & Accountability Journal, 30*(7), pp. 1481–1510.

Lambrechts, W. (2015). The Contribution of Sustainability Assessment to Policy Development in Higher Education. *Assessment & Evaluation in Higher Education, 40*(6), pp. 801–816.

Lander, E. and Castro-Gómez, S. (2011). La Colonialidad del Saber: Eurocentrismo y Ciencias Sociales. Fundación Centro de Integración, Communicación, Cultura y Sociedad (CICCUS). Haiti: Consejo Latinoamericano de Ciencias Sociales (CLASCO).

Lang, D.J., Wiek, A., Bergmann, M., Stauffacher, M., Martens, P., Moll, P., Swilling, M. and Thomas, C.J. (2012). Transdisciplinary Research in Sustainability Science: Practice, Principles, and Challenges. *Sustain Science, 7*(1), pp. 25–44.

Lao-tsu (1991). *Tao Te Ching* (Feng, G-H. and English, J. trans.). Aldershot, Hampshire: Wildwood House.

Larrinaga-Gonzalez, C. (2007). Sustainability Reporting: Insights from Neoinstitutional Theory. In J. Unerman, J. Bebbington, and B. O'Dwyer (eds.), *Sustainability Accounting and Accountability*, pp. 150–167. Oxford: Routledge.

Latour, B. (2013). *An Inquiry into Modes of Existence: An Anthropology of the Moderns*. Harvard: Harvard College.

Laughlin, R. C. (1987). Accounting Systems in Organisational Contexts: A Case for Critical Theory. *Accounting, Organizations and Society, 12,* pp. 479–502.

Leadbitter, J. (2002). PVC and Sustainability. *Progress in Polymer Science, 27,* pp. 2197–2226.

Lehman G. (2017). The language of Environmental and Social Accounting Research: The Expression of Beauty and Truth. *Critical Perspectives on Accounting, 44,* pp. 30–41.

Leopold, A. (1970). *A Sand County Almanac with Essays on Conservation from Round River.* New York: Ballantine (original work published in 1949).

Leopold, A. (1989). *A Sand County Almanac: And Sketches Here and There.* Oxford: Oxford University Press (original work published in 1949).

Levitas R. (2013). *Utopia as Method: The Imaginary Reconstruction of Society.* London: Palgrave Macmillan.

Lewis, L. and Russell, S. (2011). Permeating Boundaries: Accountability at the Nexus of Water and Climate Change. *Social and Environmental Accountability Journal, 31*(2), pp. 117–123.

Lewis, M., Dodd, J. and Emmons, R. (2009). *The Rough Guide to Vietnam.* London: Rough Guides Ltd.

Li, Y., Cheng, H., Beeton, R. J. S., Sigler, T. and Halog, A. (2016). Sustainability from a Chinese Cultural Perspective: The Implications of Harmonious Development in Environmental Management. *Environment, Development and Sustainability, 18*(3), pp. 679–696.

Liaoning EPB. (2008). Official Report of the Environmental Circumstances of Liaoning Province. Liaoning Environmental Protection Bureau, China. Retrieved from: http://218.60.146.63/hbj/UploadFile/20101027035330728. pdf

Lidköping Kommun (2001). *This is Our Future! A Vision Process Lidköping in the Future.* Lidköping: Lidköping Kommun.

Linnenluecke, M. and Griffiths, A. (2010). Corporate Sustainability and Organisational Culture. *Journal of World Business, 45,* pp. 357–366.

Linton, R. (1945). The Cultural Background of Personality. *Journal of Philosophy, 43*(4).

Locke, R., Kochan, T., Romis, M. and Qin F. (2007). Beyond Corporate Codes of Conduct: Work Organization and Labour Standards at Nike's Suppliers. *International Labour Review, 146*(1/2), pp. 21–37.

London, T. (2009). Making Better Investments at the Bottom of the Pyramid. *Harvard Business Review,* May, pp. 106–113.

Lopes, V. and Gustavo W. (2017). Milhares de Ações Sobre a Tragédia de Mariana se Arrastam na Justiça. Retrieved from: https://www.em.com.br/app/noticia/gerais/2017/08/09/interna_gerais,890448/milhares-de-acoes-sobre-a-tragedia-de-mariana-se-arrastam-na-justica.shtml

Luederitz, C., Schäpke, N., Wiek, A., Lang, D.J., Bergmann, M., Bos, J.J. and Farrelly, M. A. (2016, in press). Learning Through Evaluation — A Tentative Evaluative Scheme for Sustainability Transition Experiments. *Journal of Cleaner Production*.

Lyotard, J. F. (1993). *Le Postmoderne Explique aux Enfants Correspondance*. Paris: Centre for Digital Philosophy.

Ma, J. (2007). Ecological Civilisation is the Way Forward. China Dialogue. Retrieved from https://www.chinadialogue.net/article/show/single/en/1440-Ecological-civilisation-is-the-way-forward

MacTaggart, J. and Ullberg-Ornell, P-E. (2012). *Biosphere Innovation System-BIS*. Mariestad: Mariestad Kommun.

Madden, L. and Dell'Angelo, T. (2016). Using Photojournals to Develop Ecoliteracy in a Blended Environmental Science Course. *Journal of College Science Teaching*, *46*(1), pp. 26–31.

Makunda, C.S. (2017a). Bridging the Divide Between Problem and Solution: A Design Approach to Housing Production in Nairobi. In A. L. Bang, M. Mikkelsen, and A. Flinck (eds.). Proceedings of Cumulus REDO Conference. Paper presented at The Cumulus REDO Conference: REDO, Design School Kolding, Kolding, Denmark, 30 May — 2 June, pp. 658–661. Kolding: Design School Kolding and Cumulus International Association of Universities and Colleges of Art, Design and Media.

Makunda, C.S. (2017b). Harnessing Cultural Heritage for Locally Relevant Interior Design Solutions for New Apartments in Nairobi. Paper presented at the Nairobi Innovation Week 2017 Innovation Research Symposium: Innovating to Solve Pressing Local and Global Challenges. Nairobi: University of Nairobi.

Makunda, C.S. and Edeholt, H. (2016). How African design perspectives challenge sustainable development. In J. J. de Melo, A. Disterheft, S. Caeiro, R.F. Santos, and T. B. Ramos (eds.). Proceedings of the 22nd International Development Research Society Conference. Paper presented at the 22nd International Development Research Society (ISDRS) Conference: Rethinking Sustainability Models and Practices: Challenges for the New and Old World Contexts, Universidade Nova de Lisboa, Campus de Campolide, Lisbon, Portugal, 13–15 July, pp. 136–150. Lisbon: International Sustainable Development Research Society (ISDRS).

McDonald, S., Oates, C.J., Young, C.W. and Hwang, K. (2006). Toward Sustainable Consumption: Researching Voluntary Simplifiers. *Psychology & Marketing*, *23*(6), pp. 515–534.

McDonough, W. and Braungart, M. (2002). *Cradle to Cradle: Remaking the Way We Make Things*. New York: North Point Press.

McMichael, P. and Raynolds, L.T. (1994). Capitalism, Agriculture and World Economy. In *Capitalism and Development*, (ed. L. Sklair). London: Routledge, pp. 316–338.

McNiff, S. (ed.). (2013). *Art as Research: Opportunities and Challenges*. Bristol, Chicago: Intellect.

Medawar C. (1976). The Social Audit: A Political View. *Accounting, Organizations and Society*, 1(4), pp. 389–394.

Mendes, E.B. (2018). Interview with Andressa Schröder. (Personal Communication).

Mendes, E.B., Lima, M.S., Emerich, A.P.F. and Chilinque, T.L. (2017). Volume Morto: Performance e Corporeidade. *Fractal: Revista de Psicologia*, *29*(2), pp. 127–134.

Menon, A. and Menon, A. (1997). Enviropreneurial Marketing Strategy: The Emergence of Corporate Environmentalism as Marketing Strategy. *Journal of Marketing*, *61*(1), pp. 51–67.

Mignolo, W. (2000). La Colonialidad a lo Largo ya lo Ancho: El Hemisferio Occidental en el Horizonte Colonial de la Modernidad. In E. LANDER (comp.) La Colonialidad del Saber: Eurocentrismo y Ciencias Sociales. Perspectivas latinoamericanas. Buenos Aires. pp. 55–85. Haiti: CLACSO.

Miller, J. (2003). Daoism and Nature. In H. Selin (ed.), *Nature Across Cultures: Views of Nature and the Environment in Non-Western Cultures*, pp. 393–409. London: Kluwer Academic Publishers.

Milne M.J., Tregigda, H.M. and Walton, S. (2009). Words Not Actions! The Ideological Role of Sustainable Development Reporting. *Accounting Auditing and Accountability Journal*, 22(8), pp. 1211–1257.

Ministry of Agriculture, Forestry and Fisheries. (2006). What is "Shokuiku (Food Education)"? Retrieved from http://www.maff.go.jp/e/policies/tech_res/attach/pdf/index-1.pdf

Ministry of Justice Stockholm. (2000). SOU 2000:1 A Sustainable Democracy — Policy for Renewal in the 2000 Century. Stockholm: Ministry of Culture.

Mittelstaedt, J. and Kilbourne. W. (2006). Macromarketing as Agrology: Macromarketing Theory and the Study of the Agora. *Journal of Macromarketing*, *26*(2), pp. 131–142.

Moneva, J.M., Archel, P. and Correa, C. (2006). GRI and the Camouflaging of Corporate Unsustainability. *Accounting Forum 30*(2), pp. 121–137.

Monkhouse, L., Barnes, B. and Pham, T.S.H. (2013). Measuring Confucian Values Among East Asian consumers: A Four Country Study. *Asia Pacific Business Review, 19*(3), pp. 320–336.

Montero, M. (1998). Paradigmas, Conceptos y Relaciones Para Una Nueva Era. Cómo Pensar Las Ciencias Sociales Dese América Latina. In Seminario Las Ciencias Económicas y Sociales: Reflexiones de Fin de Siglo, Dirección de Estudios de Postgrado, Facultad de Ciencias Económicas y Sociales. Caracas: Universidad Central de Venezuela.

Mort, G.S. (2010). Sustainable Business. *Journal of World Business, 45*, pp. 323–325.

Murueva, E. (2007). Ecologitcheskie Aspekty Bugalterskogo Uchjeta: Na Primere Esnogo Sectora Economiki [Ecological Aspects of Accounting: An Example of the Forest Industry Economy]. Dissercat.com. Retrieved 12 October 2017, from http://www.dissercat.com/content/ekologicheskie-aspekty-bukhgalters kogo-ucheta-na-primere-lesnogo-sektora-ekonomiki#ixzz4r5rQBI95

Nair, A. (2005). *Sustainability of Microfinance Self Help Groups in India: Would Federating Help?* (Vol. 3516). Washington: World Bank Publications.

Nairobi City County. (2018) Nairobi City County web portal. Retrieved from http://www.nairobi.go.ke/

Naess, A. (1995). *Self-realisation: An Ecological Approach to Being in the World.* Perth: Murdoch University Press.

Neves, A.C.O., Nunes, F.P., de Carvalho, F.A. and Fernandes, G.W. (2016). Neglect of Ecosystems Services by Mining, and the Worst Environmental Disaster in Brazil. *Natureza and Conservação, 14*, pp. 24–27.

Newhouse, N. (1991). Implications of Attitude and Behavior Research for Environmental Conservation. *The Journal of Environmental Education, 22*(1), pp. 26–32.

Ngo, D.T. (2010). *Mother Goddess Religion in Vietnam.* Hanoi, Vietnam: Religion Publisher.

Nguyen, T.T., (2008). *The History of Buddhism in Vietnam.* Hanoi, Vietnam: Institute of Philosophy, Vietnamese Academy of Social Sciences. (The Council for Research in Values and Philosophy).

Nidumolu, R., Prahalad, C. K. and Rangaswami, M. R. (2009). Why Sustainability is Now the Key Driver of Innovation. *Harvard Business Review*, September, pp. 57–64.

Nicolescu, B. (2006). Transdisciplinarity, Past, Present and Future. In Bertus, H. and Reijntjes, C. (ed.). *Moving Worldviews — Reshaping Sciences, Policies and Practices for Endogenous Sustainable Development*. COMPAS Editions: Holland.

Nielsen, R.P. (2016). Action Research as an Ethics Praxis Method. *Journal of Business Ethics*, *135* (3), pp. 419–428.

Nkomo, S.M. (2011). A Post-colonial and Anti-colonial Reading of 'African' Leadership and Management in Organization Studies: Tensions, Contradictions and Possibilities. *Organization*, *18*, pp. 365–386.

Okereke, C. (2006). Global Environmental Sustainability: Intragenerational Equity and Conceptions of Justice in Multilateral Environmental Regimes. *Geoforum*, *37*(5), pp. 725–738.

Orr, D.W. (2004). *Earth in Mind: On Education, Environment, and the Human Prospect*. Washington DC: Island Press.

Oswald, J.P.F. (2014). What Does Eco-civilisation Mean? The China Story. Retrieved from http://www.thechinastory.org/2014/09/what-does-eco-civilisation-mean/

Owen D. (2008). Chronicles of Wasted Time? A Personal Reflection on the Current State of, and Future Prospects For, Social and Environmental Accounting Research. *Accounting, Auditing and Accountability Journal*, *21*(2), pp. 240–267.

Oxford University Press. (2017). Dictionary 'Evaluation'. Retrieved from https://en.oxforddictionaries.com/definition/evaluation

Palmer, M. and Finlay, V. (2003). *Faith in Conservation — New Approaches to Religions and the Environment*. Washington, DC: The World Bank.

Palmujoki, E. (2016). *Vietnam and the World: Marxist-Leninist Doctrine and the Changes in International Relations 1975-93*. Berlin: Springer.

Pan, Y. (2011). Ecological Wisdom of the Ages. The Forum on Religion and Ecology at Yale. Retrieved from http://fore.research.yale.edu/news/item/ecological-wisdom-of-the-ages/

Passmore, J. (1980). *Man's Responsibility for Nature*. London: Duckworth.

Payne, P. (2016). The Politics of Environmental Education: Critical Inquiry and Education for Sustainable Development. *The Journal of Environmental Education*, *47*(2), pp. 69–76.

Pearce, D.W. and Turner, R.K. (1989). *Economics of Natural Resources and the Environment*. Baltimore, USA: Johns Hopkins University Press.

Peattie, K. (1999). Rethinking Marketing: Shifting to a Greener Paradigm. In M. Charter and J. Polonsky (eds.), *Greener Marketing: A Global Perspective on Greening Marketing Practice*. Sheffield: Greenleaf Publishing.

Peters, M.A. (2012). Neoliberalism, Education and the Crisis of Western Capitalism. *Policy Futures in Education 10*(2), pp. 134–141.

Pham, T.H.G. (2017). Mother Nature is Not Your Slave, Bao Phunu. Retrieved from http://phunuonline.com.vn

Phan, H.L., Tran, Q.V., Ha, V.T. and Lương, N. (1991). *History of Vietnam*. Hanoi, Vietnam: University and Professional Education Publishing House.

Pickering, M. (ed.). (2008). *Research Methods for Cultural Studies*. Edinburgh: Edinburgh University Press.

PlasticFreeJuly. (2017). Plastic Free July. Retrieved from http://www.plasticfreejuly.org/

Plumwood, V. (2002). *Environmental Culture: The Ecological Crisis of Reason*. Oxon: Routledge.

Polman, P. (2017). Unilever Questions Government Cuts to Green Energy Support. Retrieved from http://www.bbc.co.uk/news/business-35010243

Porritt, J. (2007). *Capitalism as if the World Matters*. London: Earthscan.

Porter, M.E. (2008). The Five Competitive Forces that Shape Strategy. *Harvard Business Review*, January, pp. 79–93.

Porter, M.E. (1985). *Competitive Advantage: Creating and Sustaining Superior Performance*. New York: The Free Press.

Porter, M.E. and van der Linde, C. (1995a). Green and Competitive: Ending the Stalemate. *Harvard Business Review, 73*(5), pp. 120–134.

Porter, M.E. and van der Linde, C. (1995b). Towards a New Conception of the Environment: Competitiveness Relationships. *Journal of Economic Perspectives, 9*(4), pp. 97–118.

Prahalad, C.K. and Hart, S. (2002). The Fortune at the Bottom of the Pyramid. *Strategy and Business* (January), pp. 54–67.

Project Everyone. (2017). The Global Goals for Sustainable Development. Retrieved from http://www.globalgoals.org/global-goals/quality-education/

Pujari, D., Peattie, K. and Wright, G. (2004). Organisational Antecedents of Environmental Responsiveness in Industrial New Product Development. *Industrial Marketing Management, 33*(5), pp. 381–391.

Pujari, D. Wright, G. and Peattie, K. (2003). Green and Competitive: Influences on Environmental New Product Development Performance. *Journal of Business Research, 56*(8), pp. 657–671.

Purser, R.E., Park, C. and Montuori, A. (1995). Limits to Anthropocentrism: Toward an Ecocentric Organization Paradigm? *Academy of Management Review, 20*(4), pp. 1053–1089.

Quattrone, P. (2015). Governing Social Orders, Unfolding Rationality, and Jesuit Accounting Practices: A Procedural Approach to Institutional Logics. *Administrative Science Quarterly*, *60*(3), pp. 1–35.

Quayson, A. (2000). *Post-colonialism: Theory, Practice or Process*. Oxford: Polity.

Rajecki, D.W. (1982). *Attitudes: Themes and Advances*. Sineuar: Sunderland.

Ranganathan, J. (1998). *Sustainability Rulers: Measuring Corporate Environmental and Social Performance: Sustainable Enterprises Perspectives series*. Washington DC: World Resources Institute.

Raus, R. and Falkenberg, T. (2014). The Journey Towards a Teacher's Ecological self: A Case Study of a Student Teacher. *Journal of Teacher Education for Sustainability*, *16*(2), pp. 103–114.

Reddy, T.P. (2010). Distress and Deceased in Andhra Pradesh: An Analysis of Causes of Farmers' Suicide. In *Agrarian Crisis and Farmers' Suicides*, (eds. R.S. Deshpande, and S. Arora), Thousand Oaks, CA: SAGE Publications.

Revans, R. (1981). What is Action Learning? *Journal of Management Development*, *1*(3), pp. 64–75.

Revans, R.W. (2011). *ABC of Action Learning*. Farnham: Gower.

Revathi, E. and Galab, S. (2010). Economic Reforms and Regional Disparities. In (eds. R.S. Deshpande, and S. Arora), *Agrarian Crisis and Farmers' Suicides*. Thousand Oaks, CA: SAGE Publications.

Roberts, J. and Scapens, R. (1985). Accounting Systems and Systems of Accountability — Understanding Accounting Practices in Their Organizational Contexts. *Accounting Organizations and Society*, *10*(4), pp. 443–456.

Rodrigues, C. (2017). Scenic Experience Foz Afora. Espaço Cultural Municipal Sérgio Porto-RJ. Retrieved from: https://www.coletivoliquidaacao.com/ocupacao

Rognova, O. (2013). Actualnie Problemi Finansovoj Otchjetnosti [Current Problems of Financial Reporting]. CyberLeninka. Retrieved 12 October 2017 from https://cyberleninka.ru/article/n/aktualnye-problemy-finansovoy-otchetnosti

Rolston, H. (1994). *Conserving Natural Values*. New York: Columbia University Press.

Roper, J. (2007). *The United States and the Legacy of the Vietnam War*. London: Palgrave Macmillan.

RSPP (2018). Natchionalnij Registr i Biblioteka Korporativnih Nefinansovih Otchetjov [National Register and Library of Non-Financial Corporate Reports]. Retrieved from http://xn--o1aabe.xn--p1ai/simplepage/157.

Russell, S., Milne, M.J. and Dey, C. (2017). Accounts of Nature and the Nature of Accounts: Critical Reflections on Environmental Accounting and Propositions for Ecologically Informed Accounting. *Accounting, Auditing & Accountability Journal, 30*(7), pp. 1426–1458.

Said, E. (1987). *Orientalism*. London: Henley Routledge & Kegan Paul.

Sainath, P. (2009). Thirty-four Months with no Income. The Hindu. [Online], September 01. Retrieved from: http://www.thehindu.com/opinion/columns/sainath/thirtyfour-months-with-noincome/article12859.ece

Sainath, P. (2010). Our Drought Came Before Yours. The Hindu. [Online], January 05. Retrieved from: http://www.thehindu.com/opinion/columns/sainath/our-drought-came-beforeyours/article10427.ece

Sainath, P. (2011). Census Findings Point to Decade of Rural Distress. The Hindu. [Online], September 26. Retrieved from: http://www.thehindu.com/opinion/columns/sainath/census-findings-point-to-decade-of-ruraldistress/article2484996.ece

Sainath, P. (2013). Farmers' Suicide Rates Soar Above the Rest. The Hindu. [Online], May 18. Retrieved from: http://www.thehindu.com/opinion/columns/sainath/farmers-suiciderates-soar-above-the-rest/article4725101.ece.

Salemink, O. (2008). Spirits of Consumption and the Capitalist Ethic in Vietnam, In Kitiarsa, P. (ed.), *Religious Commodifications in Asia: Marketing Gods* (pp. 147–168). London and New York: Routledge.

Santos, B.d.S.N., J. Arriscado and M.P. Meneses. (2007). *Opening up the Canon of Knowledge and Recognition of Difference Another Knowledge is Possible (Vol. XIX–LXII)*. London: Verso.

Sen, A. (1989). Development as Capability Expansion. *Development Planning, 19*, pp. 41–58.

Sen, A. (1999). *Development as Freedom*. New York: Knopf.

Sengupta, A., Kannan, K., Srivastava, R., Malhotra, V. and Papola, T. (2007). Report on Conditions of Work and Promotion of Livelihoods in the Unorganised Sector. New Delhi: National Commission for Enterprises in the Unorganised Sector, Government of India.

Scardamalia, M. and Bereiter, C. (2006). Knowledge Building: Theory, Pedagogy, and Technology. In R.K. Sawyer (ed.), *The Cambridge Handbook of the Learning Sciences* (pp. 97–118). Cambridge: Cambridge University Press.

Schaefer, A. and Crane, A. (2005). Addressing Sustainability and Consumption. *Journal of Macromarketing, 25*(1), pp. 76–92.

Schäfer, M., Ohlhorst, D., Schön, S. and Kruse, S. (2010). Science for the Future: Challenges and Methods for Transdisciplinary Sustainability Research. *African Journal of Science, Technology Innovation and Development*, 2(1), pp. 114–137.

Schaltegger, S., Álvarez Etxeberria, I. and Ortas, E. (2017). Innovating Corporate Accounting and Reporting for Sustainability — Attributes and Challenges. *Sustainable Development*, 25, pp. 113–122.

Schejtman, A. and Ranaboldo, C. (2009). *El Valor Del Patrimonio Cultural. Territorios Rurales, Experiencias y Proyecciones Latinoamericanas*. Lima: Instituto de Estudios Peruanos.

Schröder, A. (2018). The Potentials of Arts-based Research for the Study of Culture. Bachmann-Medick. In J. Kugele, (ed.), *Futures of the Study of Culture*. Berlin: DeGruyter.

Schultze, U. and Stabell, C. (2004). Knowing what you Don't Know: Discourses and Contradictions in Knowledge Management Research. *Journal of Management Studies*, 41(4), pp. 549–573.

Schumacher E.F. (1973). *Small is Beautiful*. London: Abacus.

Seitz, M. and Peattie, K.L. (2004). Meeting the Closed-loop Challenge: The Case of Remanufacturing. *California Management Review*, 46(2), pp. 74–89.

Senge, P. and Carstedt, G. (2001). Innovating Our Way to the Next Industrial Revolution. *MIT Sloan Management Review*, 42, pp. 24–38.

Shah, S. and Ruparel, R. (2016). The Transformation of the Housing Finance Company of Kenya. Centre for Affordable Housing Finance in Africa (CAHF). Retrieved from http://housingfinancafrica.org

Sharma, A., Iyer, G., Mehrotra, A. and Krishnan, R. (2010). Sustainability and Business-to-business Marketing: A Framework and Implications. *Industrial Marketing Management*, 39(2), pp. 330–341.

Sharma, S., Pablo, A. and Vredenburg, H. (1999). Corporate Environmental Responsiveness Strategies. *Journal of Applied Behavioral Science*, 35(1), pp. 87–108.

Sharma, S., Starik, M. and Husted, B. (2007). *Organizations and the Sustainability mosaic: Crafting Long-term Ecological and Societal Solutions*. Cheltenham, UK: Edward Elgar.

Sharma, S. and Vredenburg, H. (1998). Proactive Corporate Environmental Strategy and the Development of Competitively Valuable Organisational Capabilities. *Strategic Management Journal*, 19(8), pp. 729–753.

Shiva, V. (1991). *The Violence of the Green Revolution: Third World Agriculture, Ecology and Politics*. London: Zed Books.

Shiva, V. (2008). *Soil Not Oil: Environmental Justice in a Time of Climate Crisis.* Cambridge, MA: South End Press.

Shrivastava, P. (1995). The Role of Corporations in Achieving Ecological Sustainability. *Academy of Management Review, 20*(4), pp. 936–960.

Sidhu, M.S. (2010). Globalisation Vis-á-vis Agrarian Crisis in India. In (eds. R. Deshpande, and S. Arora), *Agrarian Crisis and Farmer Suicides.* Thousand Oaks, CA: SAGE Publications.

Sikka, P. (2010a). Smoke and Mirrors: Corporate Social Responsibility and Tax Avoidance. *Accounting Forum, 34*(3–4), pp. 153–168.

Sikka, P. (2010b). Using the Media to Hold Accountants to Account: Some Observations. *Qualitative Research in Accounting and Management, 7*(3), pp. 270–280.

Simoën, J-C. (2013). *In Search of the Missing Civilizations.* Paris: Perrin.

Singh, P., Shahi, B. and Singh, K.M., (2016). Trends of Pulses Production, Consumption and Import in India: Current Scenario and Strategies. Retrieved from: https://mpra.ub.uni-muenchen.de/81589/1/MPRA_paper_81589.pdf

Sinnes, A.T. and Eriksen, C.C. (2016). Education for Sustainable Development and International Student Assessments: Governing Education in Times of Climate Change. *Global Policy, 7*(1), pp. 46–55.

Slater, S.F., Hult, G.T.M. and Olson, E.M. (2007). On the Importance of Matching Strategic Behaviour and Target Market Selection. *Journal of the Academy of Marketing Science, 35*(1), pp. 5–17.

Smart, B. (2010). *Consumer Society: Critical Issues and Environmental Consequences.* Newbury Park, CA: SAGE Publications.

Smith, L. (2017). Meet our reef warriors. Teaching Science. *The Journal of the Australian Science Teachers Association — Supplement Ultimate Careers,* pp. 32–37.

Snow, D.A. and Benford, R.D. (1998). Ideology, Frame Resonance, and Participant Mobilization. *International Social Movement Research, 1*(1), pp. 197–217.

Soto, D.B., Alejandro, H., Bethoven, O., Gonzalo, V.J. Marrugo, L. and Perez, M. (2009). San Basilio de Palenque, Cultura Presente Territorio Ausente. In A. R. Schejtman, Claudia (ed.), *El Valor del Patrimonio Cultural. Territorios Rurales, Experiencias y Proyecciones Latinoamericanas.* Lima: Instituto de Estudios Peruanos.

Souty, J. (2017). Do Nets Grow on Trees? Coletivo Líquida Açã. Retrieved from: https://www.coletivoliquidaacao.com/single-post/2017/07/10/Redes-adormecidas-de-Reg%C3%AAncia

Souty, J. (ed.). (2017). *Foz Afora: Residência Artística no Rio Doce. Coletivo Líquida Ação.* Sao Paulo, Brazil: Rumos Itaú Cultural.

Spangenberg, J.H. (2017). Hot Air or Comprehensive Progress? A Critical Assessment of the SDGs. *Sustainable Development*, 25, pp. 311–321.

Sponsel, L.E. and Natadecha-Sponsel, P. (2003). Buddhist Views of Nature and the Environment. In H. Selin (ed.), *Nature Across Cutures: Views of Nature and the Environment in Non-western Cultures* (pp. 351–371). London: Kluwer Academic Publishers.

Starik, M. and Marcus, A.A. (2000). Introduction to the Special Research Forum on the Management of Organizations in the Natural Environment: A Field Emerging from Multiple Paths, with Many Challenges Ahead. *Academy of Management Journal*, *43*(4), pp. 539–546.

Staubus, G.J. (2003). Accounting, Accountability and Auditing and Financial Scandals Over the Centuries. SSRN University of California, Berkeley. Retrieved from papers.ssrn.com/sol3/papers.cfm?abstract_id=1733229

Stead, W.E. and Stead, J.G. (2004). *Sustainable Strategic Management.* New York: M.E. Sharpe Inc.

Steiner, R. (2010). *Double Standard: Shell Practices in Nigeria Compared with International Standards to Prevent and Control Pipeline Oil Spills and the Deepwater Horizon Oil Spill.* Amsterdam: Friends of the Earth.

Sullivan, G. (2005). *Art Practice as Research: Inquiry in the Visual Arts.* London: SAGE Publications.

Swain, R.B. and Wallentin, F.Y. (2009). Does Microfinance Empower Women? Evidence from Self-help Groups in India. *International Review of Applied Economics*, *23*(5), pp. 541–556.

Swearer, D. (1998). Buddhism and Ecology: Challenge and Promise. *Earth Ethics*, *10*(1), pp. 19–22.

Tabala, S. (S.F.). Esta Tierra no es Mia.

Tanaka, N. and Miyoshi, M. (2012). School Lunch Program for Health Promotion Among Children in Japan. *Asia Pacific Journal of Clinical Nutrition*, *21*(1), pp. 155–158.

Taylor, S. A., Humphreys, M., Singley, R. and Hunter, G. L. (2004). Business Student Preferences: Exploring the Relative Importance of Web Management in Course Design. *Journal of Marketing Education*, 26, pp. 42–49.

Teckla, M., Gerryshom, M. and Njuguna, M. (2016). Reflections on Architectural Morphology in Nairobi, Kenya: Implications for Conservation of the Built Heritage. In A.M. Deisser and M. Njuguna (eds.), *Conservation of Natural and Cultural Heritage in Kenya: A Cross-disciplinary Approach*, (pp. 75–92). London: UCL Press.

Tennessee, U.O. (2017). Sen's Capability Approach. In J. Fieser (ed.), *Internet Encyclopedia of Philosophy*. Tennessee, USA. Retrieved from http://www. iep.utm.edu/

Thielemann, U. (2000). A Brief Theory of the Market — Ethically Focused. *International Journal of Social Economics, 27*(1), pp. 6–31.

Thomson, I. (2015). But does Sustainability Need Capitalism or an Integrated Report — A Commentary on 'The International Integrated Reporting Council: A story of failure' by Flower. J. *Critical Perspectives on Accounting, 27*, pp. 18–22.

Thomson, I., Dey, C. and Russell, S. (2015). Activism, Arenas and Accounts in Conflicts Over Tobacco Control. *Accounting, Auditing and Accountability Journal, 28* (5), pp. 809–845.

Tilbury, D. (1995). Environmental Education for Sustainability: Defining the New Focus of Environmental Education in the 1990s. *Environmental Education Research, 1*(2), pp. 195–212.

Tilbury, D. (2007). Monitoring and Evaluation During the UN Decade of Education for Sustainable Development. *Journal of Education for Sustainable Development, 1*(2), pp. 239–254.

The World Conservation Union, United Nations Environment Program & World Wide Fund for Nature. (1991). Caring for the Earth: A Strategy for Sustainable Living. Gland, Switzerland: IUCN/UNEP/WWF.

Tolstoy, L. (1985). Tom 22. Izbrannie Dnevniki 1895–1910 Sobranie Sochinenij v Dvadtcati Dvuh Tomah Volume 22. In Selected Diaries 1895–1910 Collected Works in Twenty-two volumes, p. 53. Moscow: Hudozhestvennaja literatura.

Torgerson, D. (1995). The Uncertain Quest for Sustainability: Public Discourse and the Politics of Environmentalism. In F. Fischer and M. Black (eds.), *Greening Environmental Policy: The Politics of a Sustainable Future* (pp. 3–20). New York: Palgrave Macmillan.

Tran, N.T. (1999). *The Fundamentals of Vietnamese Culture* (2nd ed.). Singapore: Education Publishing House.

Tran, Q.V., To, N.T, Nguyen, C.B., Lam, M.D. and Tran, T.A. (1998). *Cultural Establishment of Vietnam*. Singapore: Education Publisher.

Truong, H.Q. (2016). A Brief History of Vietnamese Writing. Vietnamese Typography. Retrieved from: www.vietnamesetypography.com/history

Truong, H.Q., Dinh, X.L. and Le, M.H. (2001). *Outline of the History of Vietnam* (4th ed.). Singapore: Education Publisher.

Tu, W. (1998). Beyond the Enlightenment Mentality. In M.E. Tucker and J. Berthrong (eds.), *Confucianism and Ecology — The Interrelation of Heaven, Earth and Humans* (pp. 3–21). Cambridge, MA: Harvard University Press.

Tu, W. (2001). The Ecological Turn in New Confucian Humanism: Implications for China and the World. *Daedalus: Journal of the American Academy of Arts and Sciences, 130*(4), pp. 243–264.

Tucker, V. (1999). The Myth of Development: A Critique of Eurocentric Discourse. In R. Munck and D. O'Hearn (eds.), *Critical Development Theory,* (pp. 1–26). London: Zed Books.

Turgenev, I. (1971). OTCY I DETI Daty Napisanija: 1860–1861 gg. Istochnik: Biblioteka Vsemirnoj Literatury. Serija Vtoraja. Tom 117. In Fathers and Sons Dates: 1860-1861. Moscow: Hudozhestvennaja Literatura.

Turner, B.S. and Salemink, O. (2014). *Routledge Handbook of Religions in Asia.* London: Routledge.

Ullberg-Ornell, P-E. (2014). *Enterprise Democracy.* Lidköping: Lidköping Kommun.

UNESCO. (2005). United Nations Decade of Education for Sustainable Development (2005–2014): International Implementation Scheme. Paris: UNESCO Education Sector.

United Nations. (1992). UN Conference on Environment and Development: Agenda 21. Retrieved from https://sustainabledevelopment.un.org/content/documents/Agenda21.pdf

United Nations. (2016). Sustainable Development Goals: 17 Goals to Transform Our World. Retrieved from http://www.un.org/sustainabledevelopment/cities/

United Nations Department of Economic and Social Affairs Population Division. (2015). World Population Prospects, the 2015 Revision. Retrieved from https://esa.un.org/unpd/wpp/

United Nations Educational, Scientific and Cultural Organization [UNESCO]. (2016a). Text of the Convention for the Safeguarding of the Intangible Cultural Heritage. Retrieved from http://www.unesco.org/culture/ich/en/convention

United Nations Educational, Scientific and Cultural Organization [UNESCO]. (2016b). The Ngorongoro Declaration: A Major Breakthrough for African World Heritage and Sustainable Development. Retrieved from http://whc.unesco.org/en/news/1506

United Nations Environment Programme (UNEP). (2012). Global Environmental Outlook (GEO-5); Measuring Progress, Environmental Goals, and Gaps. Nairobi: UNEP.

United Nations Millennium Ecosystem Assessment. (2005). Living Beyond Our Means: Natural Assets and Human Well-Being: Statement from the board. Retrieved from http://www.millenniumassessment.org/en/Products.Board Statement

Unruh, G. and Ettenson, R. (2010). Growing Green: Three Smart Paths to Developing Sustainable Products. *Harvard Business Review*, June, pp. 94–100.

Vaidyanathan, A. (2010). *Agricultural Growth in India: Role of Technology, Incentives, and Institutions.* Oxford: Oxford University Press.

Varadarajan, R. (2010). Strategic Marketing and Marketing Strategy: Domain, Definition, Fundamental Issues and Foundational Premises. *Journal of the Academy of Marketing Science*, *38*(2), pp. 119–140.

Varey, R.J. (2010). Marketing Means and Ends for a Sustainable Society: A Welfare Agenda for Transformative Change. *Journal of Macromarketing*, *30*(2), pp. 112–126.

Varey, R.J. (2011). A Sustainable Society Logic for Marketing. *Social Business*, *1*(1), pp. 69–83.

Vietnamese Culture. (2018). Taoism in Vietnam. Vietnam Culture. Retrieved from: http://www.vietnam-culture.com/articles-108-16/Taoism.aspx

Vietnam Embassy. (2017). Official Website, Vietnam Embassy. Retrieved from: www.vietnamembassy-usa.org

Vietnam War Reference Library. (2001). The War's Effect on the Vietnamese Land and People. Detroit, USA: The Gale Group Inc.

Vu, T.A. (2006). Worshipping the Mother Goddess. *Explorations in South East Asian Studies*, *6*(1), pp. 27–44.

Wackernagel, M. and Rees, W. (1996). *Our Ecological Footprint: Reducing Human Impact on the Earth.* British Colombia, Canada: New Society Publishers.

Whitehead, A.N. (2001 [1954]). *Dialogues of Alfred North Whitehead.* Jaffrey, New Hampshire: Godine.

Wolsey, T.D. (2015). The School Walls Teach: Student Involvement in the Green School. In T. C. Chan, E. G. Mense, K. E. Lane and M. D. Richardson (eds.), *Marketing the Green School: Form, Function, and the Future.* Hershey, PA: IGI Global.

World Bank. (2014). World Bank Global Indicators. World Bank Databank. Retrieved from http://data.worldbank.org/indicator/SP.URB.TOTL.IN.ZS

World Bank. (2015). Urban Population % of Total. Retrieved from http://data.worldbank.org/indicator/SP.URB.TOTL.IN.ZS

World Bank. (2016). Economic Update: Kenya's Economy Strong in a Challenging Global Environment. Retrieved from http://www.worldbank.org/en/country/kenya/publication/kenya-economic-update-economy-strong-challenging-global-environment

World Commission on Environment and Development: 1987. (The Brundtland Report) Our common future. New York: Oxford University Press.

World Design Summit. (2017). Designing for the Future. Retrieved from https://worlddesignsummit.com/summit-of-international-organizations/

Worldometers. (2017). Kenya Population Forecast. Retrieved from http://www.worldometers.info/world-population/kenya-population/

World Inequality Lab. (2018). World Inequality Report 2018. Retrieved from: www.WIr2018.WId.World.

WWF. (2012). Living Planet Report 2012. Gland, Switzerland: WWF.

WWF. (2016). Living Planet Report: Risk and Resilience in a New Era. Gland, Switzerland: WWF.

Xiao, H. (2006). Corporate Environmental Accounting and Reporting in China: Current Status and the Future. In S. Schaltegger, M. Bennett, and R. Burritt (eds.), *Sustainability Accounting and Reporting* (pp. 457–471). Dordrecht: Springer.

Yost, D.S., Sentner, S.M. and Forlenza-Bailey, A. (2000). An Examination of the Construct of Critical Reflection: Implications for Teacher Education Programming in the 21st Century. *Journal of Teacher Education*, *51*(1), pp. 39–49.

Young, W. and Tilley, F. (2006). Can Businesses Move Beyond Efficiency? The shift Toward Effectiveness and Equity in the Corporate Sustainability Debate. *Business Strategy and the Environment*, *15*, pp. 402–415.

Zhang, J. (2010). Probing into the Environmental Accounting Information Disclosure by Chinese companies. (Personal communication).

Zocco, D. (2009). Risk Theory and Student Course Selection. *Research in Higher Education*, *3*, pp. 1–29.

Zohar, D. and Marshall, I. (2000). *Spiritual Intelligence: The Ultimate Intelligence*. London: Bloomsbury.

Zohar, D. and Marshall, I. (2004). *Spiritual Capital: Wealth We Can Live By*. San Francisco: Berrett-Koehler.

Zooprotein.com. (2017). Organic Waste Recycling into Protein Feeds and Fertilizers. Retrieved from http://www.zooprotein.com/en.html

Index

Printed in the United States
By Bookmasters